# Communications
# in Computer and Information Science  **1343**

More information about this series at http://www.springer.com/series/7899

Jasminka Hasic Telalovic ·
Mehmed Kantardzic (Eds.)

# Mediterranean Forum – Data Science Conference

First International Conference, MeFDATA 2020
Sarajevo, Bosnia and Herzegovina, October 24, 2020
Revised Selected Papers

 Springer

*Editors*
Jasminka Hasic Telalovic ⓘ
University Sarajevo School of Science
and Technology
Sarajevo, Bosnia and Herzegovina

Mehmed Kantardzic ⓘ
University of Louisville
Louisville, KY, USA

ISSN 1865-0929        ISSN 1865-0937 (electronic)
Communications in Computer and Information Science
ISBN 978-3-030-72804-5        ISBN 978-3-030-72805-2 (eBook)
https://doi.org/10.1007/978-3-030-72805-2

This Springer imprint is published by the registered company Springer Nature Switzerland AG
The registered company address is: Gewerbestrasse 11, 6330 Cham, Switzerland

# Preface

These proceedings contain papers presented at the Mediterranean Forum – Data Science Conference (MeFDATA2020), held in Sarajevo, Bosnia and Herzegovina on October 24th, 2020. This was the fourth edition of the Mediterranean Forum and the first time that a data science conference was organized. To support the development of data science tools and applications in the region, we wanted to provide an international forum in which data science academics and practitioners can communicate their findings and ideas.

We received 26 submissions encompassing a wide range of topics in data science and its applications. The submissions were reviewed in a single-blind review process. In the end, 11 papers were accepted for publishing in these proceedings. Each paper was refereed by at least three reviewers. The final acceptance was based on these reviews and decided by the consensus of both conference chairs.

Due to the pandemic conditions, the conference was held online using the Zoom platform.

Sincere thanks go to our program committee members for referring submitted manuscripts; Naida Dupovac, Jasmina Bajramovic, Sanela Pripo, and student volunteers for help in conference planning and organization.

January 2021                                                    Jasminka Hasic Telalovic

# Organization

## Program Committee

| | |
|---|---|
| Gojko Babic | Ohio State University, USA |
| Eddie Custovic | La Trobe University, Australia |
| Amra Delic | Technical University of Vienna, Austria |
| Belma Ibrahimović | Toptal, USA |
| Ernesto Iadanza | University of Florence, Italy |
| Edouard Ivanjko | University of Zagreb, Croatia |
| Ajla Kulaglic | Istanbul Technical University, Turkey |
| Raşit Köker | Sakarya University of Applied Sciences, Turkey |
| Mariofanna Milanova | University of Arkansas at Little Rock, USA |
| Lejla Pašić | University SSST, Bosnia and Herzegovina |
| Moamer Sayed-Mouchaweh | Institut Mines-Télécom Lille Douai, France |
| Ismar Volić | Wellesley College, MA, USA |

## Conference Chairs

| | |
|---|---|
| Jasminka Hasic Telalovic | University SSST, Bosnia and Herzegovina |
| Mehmed Kantardzic | University of Louisville, USA |

# Contents

**Natural Language Processing**

# Human Behaviour and Pandemic

# Predicting the Coronavirus Spread Based on Countries' Long-Term Socio-Economic Indicators

Kemal Altwlkany$^{(\boxtimes)}$ (iD), Edina Ražanica$^{(\boxtimes)}$ (iD), Nina Mijatović$^{(\boxtimes)}$ (iD), and Amra Delić$^{(\boxtimes)}$ (iD)

Faculty of Electrical Engineering, University of Sarajevo, Sarajevo, Bosnia and Herzegovina
{kaltwlkany1,erazanica1,nmijatovic1,adelic}@etf.unsa.ba

**Abstract.** In this paper, we present an approach on how to predict the coronavirus spread per country from the country-specific socio-economic indicators. To this end, firstly, we describe in detail how the growth of COVID-19 cases can be represented with a parameterized exponential curve. Then, having collected and pre-processed various country rankings, statistics and indicators of socio-economic circumstances of a country, we constructed an adequate dataset of 116 countries. In order to predict the behavior of the coronavirus spread behavior, we employed machine learning algorithms, i.e., regression and classification approach. Since the dataset is unlabelled, we also made use of the clustering methods. In essence, the results of the regression analysis indicate a strong relationship between countries' socio-economic indicators and the behavior of the coronavirus number of novel cases. Whereas, due to the lack of labeled dataset, the classification method results in a rather poor performance.

**Keywords:** Analysis · Prediction · COVID-19 · Country rankings · Machine learning · Data mining

## 1 Introduction

The COVID-19 disease has been declared a pandemic and has spread to almost all countries in the world, with more than 16.5 million confirmed cases worldwide[1], according to the World Health Organization (WHO) [1]. It is without a doubt that the pandemic has aroused many questions and drawn attention to topics from various areas of interest (e.g., economic and social consequences, mental health perseverance, etc.). One such study, done by Bonaccorsi et al. [2], models the pandemic similar to a natural disaster to gain insight into the economic and social consequences of the restricted mobility people were subjected to in Italy, during the peak of the pandemic. An interesting work by Haushofer

---

[1] At the moment of writing this paper.

© Springer Nature Switzerland AG 2021
J. Hasic Telalovic and M. Kantardzic (Eds.): MeFDATA 2020, CCIS 1343, pp. 3–15, 2021.
https://doi.org/10.1007/978-3-030-72805-2_1

and Metcalf questions which interventions work best in a pandemic, by radically examining the rationality and effectiveness of measures such as social distancing and hand-washing promotion, i.e., whether empirical evidence exist that such measures actually reduce the transmission rate of the virus [3].

As outcomes and consequences of the pandemic have taken the effect in numerous different areas of human life, i.e., economy, society, education, the use of media and others, we were motivated to investigate yet another aspect, summarized in the following research questions:

RQ1 *Is there a relationship between the economic, social, educational indicators, and general performance statistics of individual countries, on one side, and the spread of the virus on the other side?*

RQ2 If yes; *How strong is the relationship between the virus spread and these various socio-economic indicators?*

It makes sense to assume that countries with a poor healthcare system and a low ratio of hospitals to population are more likely to suffer ample consequences caused by the pandemic. Nonetheless, it is valid to question whether a correlation can be found between the economic strength of a country and the spread of COVID-19. This question does not limit itself to only the economic strength, but can be widened to broader indicators, such as the average life expectancy in a country, population density of a country, crime rate of a country, estimated corruption perception, and many other indicators.

To answer the research questions, with the help of several machine learning algorithms we try to predict the coronavirus spread, more specifically, to estimate the parameters of the exponential function that captures cumulative number of total cases per country, based on the country specific economic, social, etc. indicators. To this end, we collected a dataset containing the required information about countries, and the timeseries representation of the coronavirus spread.

The rest of the paper is organised as follows. Section 2 presents related work, i.e., studies that try to predict the coronavirus outbreak. In Sect. 3 the method of describing the COVID-19 growth of novel cases will be elaborated and the mathematical approach used will be outlined as well. Following in the same section is the depiction of statistics and indicators used to express and describe each country. Section 4 presents the results of the applied machine learning approaches in an attempt to discover hidden connections between countries' statistics, indicators and the rate at which COVID-19 spreads. Some issues which may arise when describing the data, as well as how to detect those issues and deal with them are discussed in Sect. 5. Finally, Sect. 6 provides a summary of the results obtained and conclusions which can be derived from the procedure completed in Sect. 3, as well as provide a very clear set of goals and guidelines for potential future research.

## 2  Related Work

In terms of predicting the transmission rate of COVID-19, an interesting work by Stubinger and Schneider employed dynamic time warping to determine the relations between nations [4]. The results indicated that China was the origin of COVID-19, whereas Italy was the epicenter in Europe. Giuliani et al. created a prediction model of the COVID-19 spread in Italy primarily motivated by spatial factors, i.e. taking into account the spatial closeness of certain regions. [5]. Nesteruk attempted to predict the characteristics of the COVID-19 epidemic in its early stages, while it was only present in China [6]. Although not that successful, the approach is very interesting and uses the SIR model [7] (based on differential equations). A prediction based on intercity travel data was completed by Zhan et al. [8]. The results obtained were not coherent with other predictions and estimates during that time and as such improvement suggestions have been provided. Some previous research also incorporated government interventions into their models (Zhang et al. [9]). Elmousalami and Hassanien have used a moving average and weighted moving average model accompanied by single and triple exponential smoothing to obtain their prediction of novel cases on a worldwide level [10]. Pal et al. predict the risk category of each country using a neural network, combining COVID-19 and weather data [11].

Clearly, the hereby analyzed topic is of great importance and many researchers attempted to predict various aspects of COVID-19 (be it the transmission rate, an estimated duration of the pandemic, or repercussions on the economy or other spheres of interest). An indicator to the challenge ahead of this approach is that many have claimed dissatisfaction with their obtained results and suggested various improvements.

## 3 Methodology

### 3.1 Information Regarding COVID-19

*Coronavirus Novel Cases.* Out of a variety of different information used to describe the pandemic, we focus on the number of novel confirmed cases per country [12].

Since the COVID-19 pandemic has also raised a lot of political issues, this paper does not discuss, proclaim nor question the sovereignty or jurisdiction of certain countries. The countries listed in the dataset are used and categorized as such (e.g., Kosovo or Taiwan [1]) without the interference or modification of the list by any of the authors.

*Describing the COVID-19 Curve.* The data obtained has a timeseries representation, where values indicate the cumulative number of total cases registered up to date. The timeseries starts as of January 22$^{nd}$ and ends with July 31$^{st}$.
In order to compare the countries and to characterize the number of novel cases based on the country specific socio-economic indicators, we make use of machine learning algorithms. In order to do this, first, a prototype function to describe the number of novel cases has been selected and is introduced in Eq. 1. Here, the independent and discrete variable $t$ has a character of time and is expressed in days. Therefore, the value of the function corresponds to the number of novel cases reported daily. Parameters $a$, $b$, $c$, and $d$ are to be estimated and are

unique for every country. Ideally, these parameters will have such values that they implicitly contain information about the character of the virus spread in the corresponding country.

$$f(t) = a + b \cdot e^{c \cdot t + d} \tag{1}$$

Equation 1 can be reformulated to Eq. 2 by simply substituting $g = b \cdot e^d$, still describing an identical curve. However, even though, parameter $b$ (or $d$), in Eq. 1, might seem redundant, it is used for a rather technical purpose, as will be elaborated later.

$$f(t) = a + g \cdot e^{c \cdot t} \tag{2}$$

In order to estimate the parameters $a, b, c$ and $d$ for each country, first a curve-fitting approach is employed. The parameters are estimated by minimizing a measure of distance between the prototype function (estimated curve) and the actual timeseries data. For this purpose we used SciPy [13] library in Python, and a curve-fitting method from the *optimize* module, which implements the Levenberg-Marquardt (LM) heuristic optimization method [14–17].

*Issues with Data Fitting.* As previously mentioned, parameters $b$ and $d$ could have been aggregated into a single parameter $g$. However, empirical observations showed that better results in curve fitting were obtained by using four instead of three parameters. The reason behind might be that using a somewhat redundant parameter allowed the optimization algorithm to converge more easily, as it had four degrees of freedom. Other issues regarding the fitting process are reviewed in the Discussion section.

## 3.2   Country Statistics

The initial goal of this work was to describe countries by using information from a variety of different areas. For instance, we hypothesized that the rate of small crimes committed by citizens (normalized by the population size) would indicate the effectiveness of restrictions imposed by the law. The rationale behind such reasoning would be that a nation who respects lesser laws will tend to respect restrictions during the virus outbreak more. By lesser laws, offenses such as parking tickets or verbal disputes are considered (generally non-aggressive violations with no lethal outcomes). However, to our best knowledge, it was not possible to obtain such information reliably. The Knoema online database contains various datasets with respect to different country rankings, statistics and information on topics such as agriculture, crime statistics, the human development index, demographics, economics and other [18], but no information regarding less penalizing crimes could be found here. Moreover, obtaining such information online via other sources was also unsuccessful, as many countries do not update this data regularly [18].

Country statistics and indicators do not change rapidly, e.g., purchasing power parity, the mortality rate, or the life expectancy at birth (clearly natural catastrophes or wars are excluded from this observation), etc. To this end, it seems reasonable to consider country statistics and indicators as an average over the past 10 years. The final decision comes down to choosing the rankings, statistics and indicators used for describing the countries, and here, a compromise had to be made between data which is instinctive to be analysed, but as well as data which was possible to obtain (e.g., the lesser crimes appear reasonable, but were not possible to obtain). Hence, the following was selected, and hereby motivated:

1. *Population density:* The idea here is that a densely populated country is more susceptible to a virus spread.
2. *Gross domestic product per capita based on purchasing-power-parity in current prices:* Intuitively, a population with a higher purchasing power parity is in a better predisposition to handle restrictions imposed by the government easier, i.e., they obtain life necessities easier and can cope with less income for longer periods.
3. *Life expectancy at birth:* A country with a higher life expectancy at birth most probably owes that to a wide number of factors, e.g., high quality health-care system with a large number of hospital beds per 1000 population, etc.
4. *Unemployment rate:* On one hand, a country with a higher unemployment rate might be more likely to face a financial crisis should the restrictions be protracted. On the other hand, countries with a very low unemployment rate might find it difficult to impose radical and strict measures such as quarantine.
5. *Corruption perceptions index:* The idea here is that countries with a smaller corruption perceptions index might be less keen on obeying restrictions and following guidelines during the pandemic.
6. *Prosperity:* A more prosperous country might find it easier to support and aid their citizens financially, but also by being able to provide them with, e.g., necessary protection equipment, adequate medical treatments, etc.
7. *Human development index (HDI):* HDI can be defined in terms of three indicators for human lives (1) living a long and healthy life (2) being able to acquire knowledge, and (3) being able to have a decent life standard [18,19].
8. *Continent:* The continent of a country captures the climate type to certain extent as well as many other socio-economic indicators.
9. *Landlocked status:* Landlocked countries which depend on importing food or similar necessities may need to keep their borders open longer. Countries with sea access can transport more goods while exposing their citizens to fewer human/foreign interactions.

### 3.3   Preparing the Dataset

The parameters describing the COVID-19 novel cases $a, b, c, d$, and the country indicators are normalized using min-max normalization to a $[-0.5, 0.5]$ range. The indicators *landlocked* and *continent* are categorical variables and thus not normalized. Moreover, *prosperity* is converted into a categorical variable due to a great difference between countries with respect to this particular measurement. Hence, scaling *prosperity* using the min-max normalization would result in many countries differing only in the 2nd or 3rd decimal point even though the difference is a lot greater when considering the original data. The ordered categories of the transformed variable *prosperity* are: *very poor, poor, modest, balanced, middle, upper middle, rich, very rich*. Lastly, the GDP indicator, based on purchasing power parity, has the same scaling issue. In order to deal with this, a different approach was used, inspired by genetic algorithms [20], which are based on evolution. Interested readers can refer to roulette-wheel selection used in genetic algorithms [20]. Based on this approach, normalization of the purchasing power parity $p$ for every country was done using Eq. 3. Variable $SP$ was set to 1.5. The result of this approach was lessening the influence of countries which almost act as outliers, having incredibly high or incredibly low values of the purchasing power parity compared to the median or mode values.

$$\text{normalized}(p) = (SP - 2) + 2 \cdot (SP - 1) \cdot \frac{\text{rank}(p) - 1}{\text{number\_of\_countries} - 1} \qquad (3)$$

To summarize, the goal of this section was to illustrate how the timeseries containing the total novel cases of COVID-19 reported per country is captured with the four variables, i.e., parameters $a, b, c$ and $d$. Afterwards, nine indicators which describe economic, social and other aspects of each country are chosen, and a brief overview of the normalization methods is presented.

What further remains is to define a procedure (or to train a machine learning model) to predict the parameters of the COVID-19 curve based on the nine predictor variables.

### 3.4    Predicting the Parameters of the COVID-19 Curve

**Regression.** Since the target variables are continuous values in the $[-0.5, 0.5]$ range, they can be estimated with regression methods. To this end, again, the Sci-kit library and the *Decision Tree Regressor* module is used [21, 22]. Specifically, in order to achieve better accuracy, the Multi-output Decision Tree Regressor (MODTR) is employed. It is applicable for cases where a correlation between the target variables exists [22], as it is in our case (demonstrated in Table 1).

**Table 1.** Correlation between coefficients $a$, $b$, $c$ and $d$

|   | b | c | d |
|---|---|---|---|
| a | −0.34 | 0.44 | −0.8 |
| b |  | 0.26 | 0.17 |
| c |  |  | −0.64 |

In order to evaluate the performance of our regression model four different performance measures were used: the mean absolute error (MAE), the mean squared error (MSE), the R-squared ($R^2$) (i.e., indicating the amount of variance explained by the model), and adjusted R-squared (adj-$R^2$), all estimated with the 5-fold cross validation.

**Classification.** To further explore the relationship between the countries' socio-economic indicators and the rate of COVID-19 transmission, we investigated the approach of classifying the countries into several groups based on the similarity of the exponential growth curve, hence parameters $a$, $b$, $c$, and $d$. Then, instead of estimating the value of each parameter with regression, we try to classify countries to a set of groups (classes) and from there deduce the values of the parameters describing the exponential curve. This either means assigning a class label to each country manually or by allowing an unsupervised machine learning algorithm to do it for us, i.e., clustering.

*Labeling the Dataset by Clustering.* As clustering is an unsupervised learning approach, usually, the primary goal is a descriptive analysis [23]. In our case, clustering will be used to assign labels to each country, where a particular label would ideally represents the pace at which the virus spreads. The number of clusters is unknown, but there is a reasonable assumption that the dataset contains from three up to eight clusters, i.e., the lower bound three corresponds to the case when there are three distinct types of virus spread: slow, medium and fast; whereas the case of eight clusters is motivated by the fact that most of our predictors, as empirically evaluated, are optimally split into eight categorically distinct values. Here, we first examine clustering tendency of the data, and as a relative measure of that tendency the Hopkins statistic coefficient is used [24]. The Hopkins coefficient is determined for various subsets of the $\{a, b, c, d\}$ set, where the parameters $a$, $b$, $c$, and $d$ are normalized either with min-max (label M) or z-score (label Z) normalization. The obtained Hopkins statistic coefficients are:

$$H_{M-abcd} = 0.70 \qquad H_{Z-abcd} = 0.62 \qquad H_{M-bcd} = 0.70 \quad H_{Z-bcd} = 0.63$$
$$H_{M-bc} = \mathbf{0.78} \qquad H_{Z-bc} = 0.72 \qquad H_{M-cd} = 0.75 \quad H_{Z-cd} = 0.47$$
$$H_{M-a,be^c,e^c} = 0.75 \quad H_{Z-a,be^c,e^c} = 0.63$$

These results suggest strong clustering tendency, therefore introducing categorical description should be plausible. In the clustering analysis we compared the performance of several methods, i.e., *KMeans, KMedoids*, and a density-based approach, *DBSCAN*, and hereby we summarize the results.

The optimal number of clusters is determined using the Elbow method, and [25]. the quality of the obtained clusters is evaluated with the Silhouette Score measure. The Silhouette coefficient lies within the $[-1, 1]$ range, where 1 represents an ideal case with compact and separated clusters.

KMeans resulted in unacceptable clusters as it categorized all countries to a single cluster. KMedoids gave interesting results, while density based clustering algorithms have been shown to not perform well on this type of data.

Empirically analysing KMedoids algorithm, for $N \in [3, 8]$, the best clustering quality criteria, hereby, was obtained for $N = 4$, and with $\{b, c\}$ subset of parameters, hence splitting the data into four clusters. The number of countries per cluster is 4, 49, 32, and 8, respectively. Therefore, each country is labeled with one of the four cluster labels, as follows:

- Cluster 0 - Extremely fast virus spread.
- Cluster 1 - Slow virus spread;
- Cluster 2 - Medium virus spread;
- Cluster 3 - Fast virus spread with two exceptions found (Trinidad and Tobago, and Uganda which are characterized with slow virus spread) (Table 2);

**Table 2.** Results of applying the KMedoids algorithm

| Cluster 0 | Cluster 1 | Cluster 2 | Cluster 3 |
| --- | --- | --- | --- |
| China | Afghanistan | Austria | Germany |
| Iran | Albania | Bahrain | Netherlands |
| Italy | Algeria | Belgium | Switzerland |
| Venezuela | Argentina | Brazil | Trinidad and Tobago |
| | Armenia | Bulgaria | Turkey |
| | Australia | Chile | Uganda |
| | Azerbaijan | Czech Republic | United Kingdom |
| | +42 Countries | +25 Countries | United States |

Finally, after labeling our data with the previously obtained clusters, the final goal is to predict to which cluster a country belongs based on its socio-economic indicators. The classifier used is the Naive Bayes algorithm (NB) also from the Sci-kit library [21, 22]. NB was also evaluated against the KNN algorithm, but in this work, we do not present the results of comparison due to the limitations on paper length.

NB is evaluated with a hold-out approach, with 65% to 35% split on training and test sets. The evaluation metrics used are precision, recall, f1-score per class (cluster) and in average, and accuracy in average over all the classes.

## 4   Results

In this section, we present the results of our regression and classification models. The results were obtained while running experiments multiple times and averaging the results from single runs.

*Regression.* Table 3 provides performance measures obtained when using the MODTR model, whereas Table 4 provides the same performance measures, but when four independent regression trees were trained. In the latter approach, each tree is used to predict one parameter. It is evident from Tables 3 and 4 that using the MODTR instead of separate regression trees, in average, results in a better tree in terms of the MAE, $R^2$ and adj-$R^2$ measures, while MSE score is actually lower for the multiple-tree prediction approach.

**Table 3.** Performance measure of the Multi-output decission tree regressor

|  | MAE | MSE | $R^2$ | adj-$R^2$ |
|---|---|---|---|---|
| Value | 0.117 | 0.097 | 0.728 | 0.679 |

**Table 4.** Performance measures of four independent regression trees which predict variables *a,b,c* and *d* distinctly.

|  | MAE | MSE | $R^2$ | adj-$R^2$ |
|---|---|---|---|---|
| a | 0.208 | 0.104 | 0.635 | 0.519 |
| b | 0.192 | 0.076 | 0.656 | 0.719 |
| c | 0.159 | 0.052 | 0.777 | 0.751 |
| d | 0.114 | 0.098 | 0.668 | 0.633 |

## 4.1   Classification

The results of the classification are demonstrated in Table 5. The performance of the classifier, in terms of precision, recall and f1-score appear to be quite satisfying, if we consider the weighted average scores. However, this is not really the case. The data are extremely unbalanced over the four clusters, where cluster 1 itself contains more than half the data, while clusters 3 and 0 contain 2 and 1 data instances respectively, in the test sample. Therefore, even though the overall results appear to be satisfying, not possessing enough instances in each class makes the classifier unreliable. In fact, precision, recall and f1-score for cluster 3 and 0 are equal to 0.00 in both cases, exactly due to this class-imbalance, hence the classifier simply cannot learn to assign any data instances to these clusters. The macro-average scores are a lot better indicator of this downside, compared to the weighted average performance measure.

The results indicate that attempting to predict the parameters which describe the COVID-19 curve using regression can be done successfully using a MODTR. An attempt on classifying the data using class labels generated via clustering does not result with acceptable results, mainly due to the class imbalance and the small number of instances in the test set. Given the strong correlation which was shown earlier, satisfactory results obtained using regression seem reasonable. The following chapter provides a verbose discussion related to the obtained results.

**Table 5.** Performance measures of NB classifier. Classification of clusters obtained using spectral clustering, 4 clusters.

|              | Precision | Recall | F1-score | Support |
|--------------|-----------|--------|----------|---------|
| Cluster 0    | 0.00      | 0.00   | 0.00     | 1       |
| Cluster 1    | 0.77      | 0.87   | 0.82     | 23      |
| Cluster 2    | 0.73      | 0.73   | 0.73     | 15      |
| Cluster 3    | 0.00      | 0.00   | 0.00     | 2       |
| Accuracy     |           |        | 0.76     | 41      |
| Macro avg    | 0.38      | 0.40   | 0.39     | 41      |
| Weighted avg | 0.70      | 0.76   | 0.73     | 41      |

## 5    Discussion

### 5.1    Issues with Data Fitting

Although pretty straightforward, the process of fitting the data cannot be done by simply attempting to fit the prototype function using the raw unprocessed data. Addressing the issues which occur while attempting to do so are crucial to understanding the character of the virus and the nature of the problem at hand.

- *Inconsistent data.* It may happen that some countries report official data which is not consistent (or that the dataset contains an error). Namely, the cumulative amount of novel cases reported can only be monotonically non-decreasing. As such, fitting the data with an exponential function will be troublesome.
- *Lack of cases.* In countries where not enough cases were registered it might not be possible to find an exponential curve with satisfying results. The LM will not "complain", it will find a curve which fits the data best, but the parameters of such a fit cannot be interpreted the same way as they can for good fits.
- *Data not updated regularly.* Somewhat related to the previous two issues is an anomaly of the number of novel cases when countries did not update them regularly. This causes the data to stagnate at a certain number of cases and then instantaneously jump to a larger number. The issue at hand here is that perhaps more frequent updates/checks on new cases might have been necessary, which as a result could have straightened out the clustered number of cases.
- *Outbreak has not begun within 75 d as of January 22$^{nd}$.* Despite the virus outbreak occurring mainly in March, many countries have prolonged the outbreak. To resolve this issue in our analysis, the data containing no cases in the beginning is simply disregarded, since fitting such a set of points would result in a bad fit.
- *Spreading is not of exponential character.* Regardless of the assumptions that COVID-19 spreads with an exponential character, this may not necessarily

be true for all countries. Lots of countries were found with an almost perfect fit, but there are quite a few cases where this does not hold. Uruguay appears to be a good example, as the data appears to be more appropriate for a linear fit.

Given all the preceding notices, 116 countries were compatible for further analysis, and the rest were disregarded from our dataset. A small percentage of the disregarded countries actually reported a significant number of COVID-19 infections, whereas most of the disregarded countries simply did not have a high rate of COVID-19 cases.

# 6  Conclusion

In this paper, the goal was to investigate whether it is possible to predict coronavirus spread per country, from the socio-economic indicators of the respective country. In order to do this, we have first collected data for each country, containing the cumulative number of cases as of January $22^{nd}$, up to July $31^{st}$, however, to make sure we isolate the effect of the restriction measures introduced in each country, we focus on the first 75 d from the first reported case. Then, we propose an approach to capture the collected timeseries data with the help of four parameters ($a$, $b$, $c$, and $d$) that determine the shape of the exponential curve describing the behavior of the coronavirus spread. After purging the dataset, we were left with the 116 countries, for which the analysis is presented in this paper. In the next step, we collected and pre-processed the socio-economic indicators for each of these countries. To answer our research questions, and predict the parameters describing the exponential curve, we did a regression analysis. Moreover, as we were interested to further understand the topic at hand, we decided to try to classify each country based on its socio-economic indicators, where the target class would be the rate of the coronavirus spread. However, the collected dataset was not labeled with a type of the rate of the coronavirus spread, and to overcome this limitation, we labeled the data with the KMedoids clustering algorithm.

In essence, the results of the regression analysis are overall satisfactory, and to our amazement, having a 70% of the variance (also in average) in the predicted curve parameters explained by the predictors, i.e., the socio-economic indicators of countries, is more than we initially expected. In contrast to this, the performance of the classification approach was at least inadequate. Here, the results are highly dependent upon the clusters to which the data is assigned. Since the clustering analysis, for the four curve parameters, did not result in comprehensive clusters, we can hardly assume that classification would in fact be of use.

Due to the obtained results, it is important to reflect on the limitations of the presented study. First of all, the dataset per country and its volume are restrictive and hinder the power of machine learning methods. Hence, in future work, it would be of great importance to investigate this topic with instances representing regions of cohesive socio-economic circumstances. It is obvious how this increases the complexity of the study, even simply by the process of collecting the adequate data. Moreover, in this paper, we have mostly presented the prediction power of the investigated methods, while we did not explain nor elaborate on the importance of the predictors individually, which would clearly be of use to better understand the phenomena. Finally, labeling the dataset is yet another challenge to be faced. Hereby, we presented a clustering approach, which showed to be inefficient. Another approach would be to have a set of experts manually label the dataset, or at least a part of the dataset from where we would be able to assign labels with the help of machine learning algorithms to the unlabeled data. Whatever the approach is to be undertaken, it is indeed a great challenge, especially if we would consider analysing regions, and not countries.

To summarize, throughout the entire scope of the paper, an attempt of using different approaches from various areas of data science, machine learning and optimization was made. As such, unique performance measures were suggested, different regression/classifying algorithms were used, comparisons were drawn between multi-output and individual regression trees and approaches inspired from optimization theory were used when normalizing the data. The obtained results are intriguing, and there is a lot that can be further explored. In the next step, one could look at the effects of different measures imposed by each country and their impact on the coronavirus spread, but also with respect to these various socio-economic indicators. Clearly, having that in many cases the spread would not be further captured by an exponential curve, different types of analysis could be used (e.g., time-series analysis). This type of analysis would also be beneficial even after the restrictions were canceled, as we strongly believe that people's behavior is quite dependent from the cultural and socio-economic characteristics of a country.

# References

1. World Health Organization - COVID-19 Situation Reports. Situation Report 191. https://www.who.int/emergencies/diseases/novel-coronavirus-2019/situation-reports/. Accessed 30 June 2020
2. Bonaccorsi, G., et al.: Economic and social consequences of human mobility restrictions under covid-19. Proc. Nat. Acad. Sci. **117**(27), 15530–15535 (2020)
3. Haushofer, J., Metcalf, C.J.E.: Which interventions work best in a pandemic? Science **368**(6495), 1063–1065 (2020)
4. Stübinger, J., Schneider, L.: Epidemiology of coronavirus covid-19: forecasting the future incidence in different countries. In: Healthcare, vol. 8, p. 99. Multidisciplinary Digital Publishing Institute (2020)
5. Giuliani, D., Dickson, M.M., Espa, G., Santi, F.: Modelling and predicting the spatio-temporal spread of coronavirus disease 2019 (covid-19) in Italy. Available at SSRN 3559569 (2020)
6. Nesteruk, I.: Statistics-based predictions of coronavirus epidemic spreading in mainland china. Innovation Biosystem Bioengineering, vol. 4 (2020)
7. Nesteruk, I.: Statistics based models for the dynamics of chernivtsi children disease. Res. Bull. Nat. Tech. Univ. Ukraine Kyiv Polytech. Inst. **5**, 26–34 (2017)

8. Zhan, C., Tse, C., Fu, Y., Lai, Z., Zhang, H.: Modelling and prediction of the: coronavirus disease spreading in china incorporating human migration data. Available at SSRN, vol. 3546051, p. 2020 (2019)
9. Zhang, X., Ma, R., Wang, L.: Predicting turning point, duration and attack rate of covid-19 outbreaks in major western countries. Chaos, Solitons Fractals **135**, 109829 (2020)
10. Elmousalami, H.H., Hassanien, A.E.: Day level forecasting for coronavirus disease (covid-19) spread: analysis, modeling and recommendations. arXiv preprint arXiv:2003.07778 (2020)
11. Pal, R., Sekh, A.A., Kar, S., Prasad, D.K.: Neural network based country wise risk prediction of covid-19. arXiv preprint arXiv:2004.00959 (2020)
12. The Humanitarian Data Exchange - Open source data. Novel Coronavirus (COVID-19) Cases Data. https://data.humdata.org/. Accessed 31 July 2020
13. Virtanen, P., et al.: SciPy 1.0: fundamental algorithms for scientific computing in python. Nat. Methods **17**, 261–272 (2020)
14. Levenberg, K.: A method for the solution of certain non-linear problems in least squares. Q. Appl. Math. **2**(2), 164–168 (1944)
15. Gavin, H.: The levenberg-marquardt method for nonlinear least squares curve-fitting problems. Dept. Civ. Environ. Eng. Duke Univ. **28**, 1–5 (2011)
16. Roweis, S.: Levenberg-marquardt optimization. University Of Toronto, Notes (1996)
17. Ranganathan, A.: The levenberg-marquardt algorithm. Tutoral LM Algorithm **11**(1), 101–110 (2004)
18. Knoema website. Free data, statistics, analysis, visualization and sharing. https://knoema.com/. Accessed 25 July 2020
19. Sagar, A.D., Najam, A.: The human development index: a critical review. Ecol. Econ. **25**(3), 249–264 (1998)
20. Kramer, O.: Genetic Algorithm Essentials. SCI, vol. 679. Springer, Cham (2017). https://doi.org/10.1007/978-3-319-52156-5
21. Pedregosa, F., et al.: Scikit-learn: machine learning in Python. J. Mach. Learn. Res. **12**, 2825–2830 (2011)
22. Buitinck, L., et al.: API design for machine learning software: experiences from the scikit-learn project. In: ECML PKDD Workshop: Languages for Data Mining and Machine Learning, pp. 108–122 (2013)
23. Rokach, L., Maimon, O.: Clustering methods. In: Maimon, O., Rokach, L. (eds) Data Mining and Knowledge Discovery Handbook, pp. 321–352. Springer, Boston, MA. https://doi.org/10.1007/0-387-25465-X_15
24. Banerjee, A., Dave, R.N.: Validating clusters using the hopkins statistic. In: 2004 IEEE International conference on fuzzy systems (IEEE Cat. No. 04CH37542), vol. 1, pp. 149–153. IEEE (2004)
25. Syakur, M., Khotimah, B., Rochman, E., Satoto, B.: Integration k-means clustering method and elbow method for identification of the best customer profile cluster. In: IOP Conference Series: Materials Science and Engineering, vol. 336, p. 012017. IOP Publishing (2018)

# Statistical Predictive Analysis of Tobacco Consumption Among Different Age Groups Using Online Survey

Rijad (Rick) Sarić[(✉)], Viet Duc (Harvey) Nguyen, and Edhem (Eddie) Čustović

La Trobe University, LTU, School of Engineering and Mathematical Sciences,
La Trobe Innovation and Entrepreneurship Foundry, LIEF, Melbourne, Australia
{r.saric,harvey.nguyen,e.custovic}@latrobe.edu.au

**Abstract.** Owing to the significant increase of tobacco products, smoking becomes the most common risk factor responsible for causing chronic disease such as lung cancer, pulmonary and coronary artery diseases. This research study focuses on extensive data analysis throughout four different hypotheses regarding tobacco consumption among different age groups and professions. Before hypotheses testing, an online survey is made to collect raw data about cigarette smoking. Afterwards, the final processed dataset is created, and some useful probability statements are derived and calculated by applying well-known statistical methods. Finally, the first three hypotheses are tested using IBM SPSS software environment including one-way Analysis of Variance (ANOVA) tool and Two-Tailed t-test. The final hypothesis is examined using the correlation matrix and developing a regression data model based on a linear correlation between independent variables of interest.

**Keywords:** Data analysis · Tobacco consumption · Survey · Two-tailed t-test · ANOVA

## 1 Introduction

Tobacco is one of the most popular smoking products and, yet tobacco consumption has leading preventable health risks. Nearly 6.5 trillion tobacco products are sold every year with around 1 billion cigarette smokers in the world [1]. Although a series of tobacco control measures are well established in Australia, the rapid population growth has in turn caused an increase of smokers including a large number of tobacco related diseases [2]. In general, tobacco is consumed by the ingestion of smoke, as the result of burning tobacco. Tobacco can be smoked by rolling into cigarettes to inhale its smoke or releasing the smoke using pipes and cigars. The smoke from burning tobacco triggers chemical reactions inside the human body and brain, dopamine and endorphins are released [3], creating a feeling of pleasure. However, in the early 1950s, scientists discovered the health impacts of smoking tobacco [4].

© Springer Nature Switzerland AG 2021
J. Hasic Telalovic and M. Kantardzic (Eds.): MeFDATA 2020, CCIS 1343, pp. 16–38, 2021.
https://doi.org/10.1007/978-3-030-72805-2_2

Tobacco smoking can be tracked back around 5000–3000 BC in Mesoamerica and South America [5]. By World War I, the cigarettes were becoming popular after widespread popularity in the USA [4]. In developed countries, cigarette consumption becomes the major cause of premature deaths. The main causes behind premature deaths were found to be respiratory diseases, cancer, and cardiovascular disease. In Australia, it led to almost 18,800 deaths in 2011 [6] and more than 400,000 deaths in the USA in the same year [7]. For instance, smoking is associated with 28% of lung cancer deaths, 37% of vascular diseases deaths and 26% of respiratory diseases deaths in the USA [6]. The major cause of death by smoking are coronary heart disease and stroke, cancers of the lung and upper airways, chronic obstructive pulmonary disease, and miscarriage and underdevelopment of fetus [8].

Dopamine and endorphins, triggered by nicotine, turn smokers into addicts of tobacco products. A study conducted by Ribeiro *et al.* [9] shows that even the lowest income smokers tend to substitute basic human needs with cigarettes, spending a significant amount of their income. Low-income Egyptians tend to use more than 10% of household expenses on tobacco products [10]. Moreover, a Chinese study revealed that tobacco consumers spend up to 60% of their salary and Philippines smoker families tend to spend 20% of household income on cigarettes [10]. Hence, cigarette or tobacco-related goods are addictive products which can make smokers partially sacrifice basic human needs to satisfy their craving. While this has been highly profitable for the tobacco industry, it has been a significant concern for global health authorities.

Research conducted by the WHO reveals that more than half of smokers in the world are young people of age greater than 18 and less than 25. This study also found that 90% of 600 million smokers of South-East Asia and Pacific market started smoking from the age under 18 [12]. The tobacco industry released flavored cigarettes which may appeal more to young smokers. The US Centers for Disease Control and Prevention showed that 67% of high school students and 49% of secondary school students used flavored tobacco products [13, 14]. Additionally, flavored tobacco products tend to have a higher price than regular tobacco products [15] which indicates students are likely to spend more money on cigarettes than other age groups.

WHO also indicates that in 2010, there were more male than female smokers in the world. Approximately 40% of the male population are smokers, while only 9% of the female population tends to consume tobacco products [16]. A research study published in 2012 found that low and middle-income countries have a higher rate of male smokers than female smokers [17], compared to WHO's statistics published in 2010. It can be seen that males in low-middle income countries are more likely to use tobacco products than females. However, women tend to use low nicotine and tar products, including low nicotine and tar cigarette, e-cigarette, flavored cigarette [18].

Most of the previous research studies have applied cross-sectional data analysis by analyzing data collected at a one-time instance. However, hypothesis testing is the most desirable way of examining data gathered using the survey. For example, Chen *et al.* [19] analyzed the data about more than 1,500 US adults to test the accuracy of the hypothesis of whether common smoker has a higher risk of developing lung cancer. Similarly, Zhang *et al.* [20] proposed decision tree-based data model to predict the behavior of the daily smoker using the data collected by the center for diseases control and prevention located in China.

This research study aims to create useful statistical knowledge that could be applied to track daily tobacco consummation among different age groups and professions. The statistical findings obtained in this research may help organizations combat tobacco consumption in terms of the most critical age groups of smokers including the main reasons behind their cigarette smoking addiction. Additionally, such statistical data can assist tobacco producers in identifying their sales and advertising actions to the smoking groups of interest.

## 2  Data Collection and Analysis

Making statistical decisions using collected data is realized using hypothesis testing. A hypothesis represents the specific, clear, testable, and formal statement (claim) connected to the possible research findings based on the particular nature of the considered population. Therefore, prior to data collection, it is necessary to state a certain statistical hypothesis. In this research, the four testable hypotheses regarding tobacco consumption are made. The first hypothesis is related to tobacco consumption by university students. Commonly students due to their stressful way of life tend to smoke more than others with different professions. As a result, students spend more money on tobacco products. The first hypothesis statement is that students spend more money on cigarettes in comparison to others with different professions. The emerging COVID-19 pandemic situation has caused a negative impact on the global economic outlook. Hence, in such situation, it can be assumed that not many smokers in the world can afford to spend more than a couple of hundred of Australian Dollars (AUD) on tobacco products every month. The second claim is that smokers do not spend in excess of $200 AUD on tobacco products per month. The next hypothesis includes cigarette consumption by gender. The third hypothesis claims that males consume more tobacco products than females. The final hypothesis includes smokers' favorite cigarette brands. Daily consumption is measured in the number of cigarette packs and average tobacco product expenses of a single smoker. The final hypothesis is that smokers who consume more packs of a certain cigarette brand are going to have higher expenditure.

In general, survey research is used to collect data by asking targeted questions to individuals or groups and afterwards observing their response. Surveys as a tool can be applied in both qualitative and quantitative research studies by utilizing different methods and instrumentation. This type of research methodology is usually applied in social and psychological science to describe and explore human behavior. The primary purpose of survey research is to quickly collect data describing the characteristics of relatively large samples. The surveys are commonly published online through social media or provided via e-mail. They can be anonymous, allowing research participants to fulfil all required fields with different opinions [21]. Therefore, data collection in this research study is completed by posting an online Google Forms survey with twelve questions regarding the consumption of tobacco products. This survey was available for a period of two weeks and it contained questions such as age, gender, profession, average expenses on tobacco products, daily consumption, smoking reason, health issues and reason for the possibility of ceasing smoking cigarettes. It was published online and shared through known social networks such as Facebook, Twitter, Instagram, and LinkedIn. Table 1 represents the survey questions and range of possible answers.

**Table 1.** The detailed description of attributes and possible outcomes of a survey created for statistical research about tobacco consumption

| Attribute | Description |
| --- | --- |
| Age | Participant's age (from 18 to 60) |
| Gender | Participant's gender (female or male) |
| Marital-status | Participant's marital status (single, in relationship, married, and divorced) |
| Profession | Participant's profession (student, professor, waiter, architect engineer, other-please specify) |
| Smoker | Does the participant smoke cigarettes (yes or no) |
| Average expenses | If participant is smoker – what are monthly expenses on cigarettes in $AUD (10, 20, 50, 100, 200, 500, other-please specify) |
| Packs | If participant is smoker – what is daily consumption of cigarette |
| Consumption | Packs (1, 2, 3, 4, 5, other - please specify) |
| Tobacco-brand | If participant is smoker – what is favorite tobacco brand (Marlboro, Chesterfield, Winfield, longbeach, Eve, Lucky Strike, other - please specify) |
| Smoke-reason | If participant is smoker – what is the reason behind smoking (stress, depression, fun, social situations, addiction, other - please specify) |
| Health-issue | If participant is smoker – do participant have any health issues (yes or no) |
| Quitting | If participant is smoker – do participant think about ceasing to smoke (yes or no) |
| Quitting-reason | If participant is smoker – what is the reason behind wishing to stop smoking (high cigarette price, health issues, social issues, other - please specify) |

Following, the collected data obtained from the online survey. Out of 108 collected answers, 99 were fully completed and further used for statistical analysis and hypothesis testing. Figure 1 shows the percentage of smokers and non-smokers where approximately 55% of participants are smokers, 45% are non-smokers. In Fig. 2, the percentage of smokers and non-smokers grouped by male and female genders is represented. It is noticeable that 32% are male, while 22% are female smokers including 24% male and 21% female non-smokers.

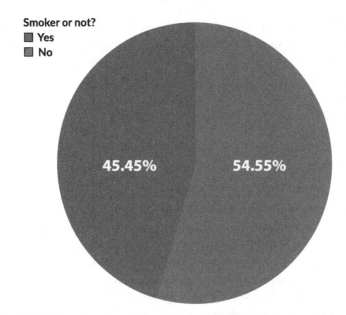

**Fig. 1.** The total number of smokers and non-smokers in the considered data sample.

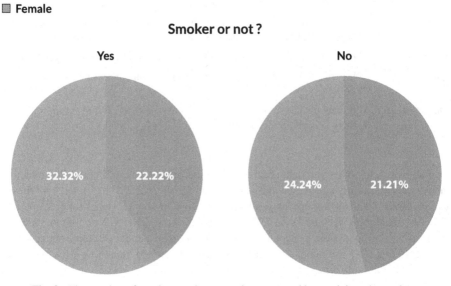

**Fig. 2.** The number of smokers and non-smokers grouped by participant's gender.

Furthermore, statistical parameters such as mean, standard deviation, median, mode, variance etc. are represented for particular attributes of the collected data (see Table 2).

**Table 2.** The detailed statistics regarding the age and gender of both smoker and non-smoker participants of the created survey.

| Age of all survey participants | | |
|---|---|---|
| N | Valid | 99 |
| | Missing | 0 |
| Mean | | 28.1010 |
| Std. Error of Mean | | 1.29717 |
| Median | | 22.0000 |
| Mode | | 21.00 |
| Std. Deviation | | 12.90665 |
| Variance | | 166.582 |
| Range | | 42.00 |
| Minimum | | 18.00 |
| Maximum | | 60.00 |
| Percentiles | 25 | 21.0000 |
| | 50 | 22.0000 |
| | 75 | 37.0000 |

As for the smokers, the mean age is 30.29 with the standard error of mean equal to 1.78. The median is equal to 22 while the mode is 21. The standard deviation has the value of 13.14. Also, the youngest smoker is 18 years old, while the oldest one is 60 years old.

| Gender of all survey participants | | | | | |
|---|---|---|---|---|---|
| | | Frequency | Percent | Valid percent | Cumulative percent |
| Valid | Male | 43 | 43.4 | 43.4 | 43.4 |
| | Female | 56 | 56.6 | 56.6 | 100.0 |
| | Total | 99 | 100.0 | 100.0 | |

From the table above it is visible that there were more females than males that took our survey. Precisely, in percentage, there was 56.6% females while there were 43.4% males.

# 3  Probability Statistics

Before the commencement of probability statements and hypothesis testing, it is necessary to identify extremely low or high values in the collected data known as outliers. This observation that differs from overall data pattern can be caused by human, instrument, data processing, experimental or intentional errors. The most common way to illustrate the distribution of collected data is a box plot. It helps in the identification of outliers and their values by displaying the distribution of five numbers namely minimum, first quartile (Q1), median, third quartile (Q3) and maximum. Figure 3 shows the boxplot which covers collected data arranged by age of smoker or nonsmoker participant attribute including identified outliers.

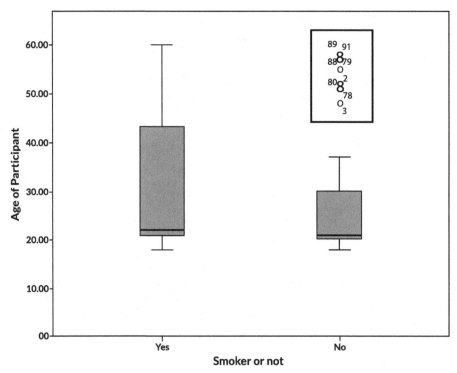

**Fig. 3.** The boxplot representing the normal distribution of collected data including error percentage of data.

The boxplot shows that there are eight outliers in nonsmoker section. To correct this, the ages of these participants need to be swapped with mean participant's age (27.67) plus two standard deviations (2 * 12.62) which equals to 53. Identified outliers disappear after swapping the actual age values at lines 3, 78, 2, 80, 88, 79, 89, 91 with calculated value 53.

Firstly, Empirical rule is applied to find some useful probabilities. This statistical rule indicates that almost all data are available within three standard deviations of the mean (average). In Fig. 4, the histogram of collected data organized by age attribute is represented, whereas Fig. 5 illustrate the Gaussian bell function of the data distribution with respect to the age of the participant.

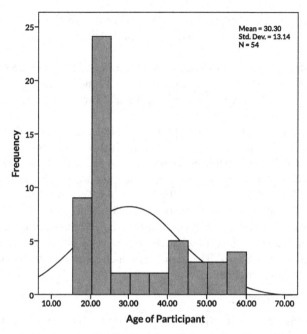

**Fig. 4.** Histogram indicating age distribution of participants together with mean and standard deviation.

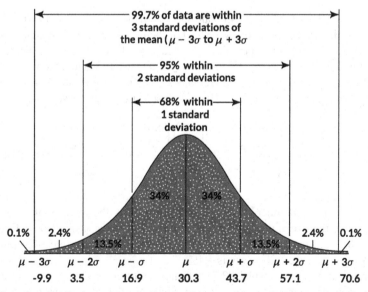

**Fig. 5.** Gaussian distribution of age attribute within one, two and three standard deviations of mean.

The empirical rule could provide more accurate approximation of smoker's age. Hence, it is possible to estimate some useful probabilities such as

- The probability that the age of smoker is between 17 and 57 years is 81,5%,
- The probability that the age of smoker is between 30 and 57 years is 47.5%,
- The probability that the age of smoker is between 3.5 and 57 years is 95%.

Based on the Gaussian data distribution of participant's age, the most useful estimation is that 68% of the smokers are between the ages of 17 and 44.

It is known that 34.6% of smokers find Marlboro as their favorite cigarette brand. Oppositely, it means that 65.4% of smokers do not choose Marlboro as their favorite cigarette brand. Out of these smokers, 40.7% are male. The probability that smoker is male, given that his favorite brand is Marlboro is 48%. The probability that smoker is male, given that his favorite brand is not Marlboro is 37%. The probability that given participant is male and his favorite cigarette brand is Marlboro is calculated in (1) by applying well-known Bayes' Theorem [22] which represents the standard way of calculating a conditional probability.

$$P(A|B) = \frac{P(B|A) * P(A)}{P(B|A) * P(A) + P(B|C) * P(C)} = \frac{0.48 * 0.346}{0.48 * 0.346 + 0.37 * 0.654} = 0.41 \tag{1}$$

It can be seen that 40.7% Marlboro smokers who participated in the created survey are males.

In the created dataset, there are 43 male and 56 female participants. If 20 participants are randomly chosen, the probability that 10 of them are male can be calculated using Hypergeometric distribution. In the case of calculating the probability of gender of the participant in the created survey, two groups of interests are made. One with a sample size of 43 (males) and another of 56 (females). The size of the sample is 20, and the probability that participant is male in 10 random selection is expressed by (2)

$$P(x = 10) = \frac{\binom{r}{x}\binom{N-r}{n-x}}{\binom{N}{n}} = \frac{\binom{43}{10}\binom{56}{10}}{\binom{99}{20}} = \frac{\frac{43!}{10!33!} \cdot \frac{56!}{10!46!}}{\frac{99!}{20!79!}} = 0.014 \tag{2}$$

where $r$ indicates the number of male participants, $N$ is the total number of survey participants, $n$ is the size of the sample, and $x$ takes on the values from 1 to 10. Evidently, the probability that ten participants are male in case of random selection among 20 participants from the created survey is 1.4%.

According to the survey results, 82.7% of smoker participants have never experienced any health issue. In a random sample of 10 participants, we want to find the probability that exactly 7 of them have never experienced any health issue. The easiest way to calculate this probability is by using the Binomial distribution. Calculating probability using binomial distribution is useful when observing a certain number of successful outcomes in a predefined number of trials. The calculation of binomial distribution on the collected data is expressed by (3)

$$P(x = 7) = \binom{10}{7} 0.827^7 0.173^3 = \frac{10!}{7!3!} 0.827^7 0.173^3 = 0.16 \tag{3}$$

where 10 is the total number of trials, 0.173 indicates the probability of failure, 0.827 indicates the probability of success and $x$ number of smokers who have never experienced any health issue. The probability that 7 of 10 smokers in this dataset have never any health issue is 16.4%.

Majority of smoker participants consume one tobacco packet per day on average. We are interested to determine the probability that single smoker consumes two tobacco packets per day using the Poisson distribution. This type of statistical distribution is particularly useful when calculating the probability of event occurrence in a specified period of time. The calculation is expressed in (4)

$$P(x = 2) = \frac{e^{-\lambda}\lambda^x}{x!} = \frac{2.718^{-1}1^2}{2!} = 0.18 \tag{4}$$

where $x$ indicates the number of tobacco packs being smoked and $\lambda$ stands for a total number of events that may occur. It is noticeable that the probability of consuming two tobacco packs per day by smokers in the created dataset is 18%.

Considering collected data, it is known that mean average for spending money on tobacco per month is $104 with a standard deviation of 47. It will be useful to determine the probability that smokers spend more than $150 monthly (specific area under normal data distribution od data). Finding this probability can be accomplished by calculating z-score as in (5)

$$Z = \frac{x - \mu}{\sigma} = \frac{150 - 104}{47} = 0.98 \tag{5}$$

where $x$ indicates average spending per month that needs to be tested, $\mu$ average spending of $104 as indicated in dataset and $\sigma$ represents calculated standard deviation. Hence, probability $P(x > 150) = P(Z > 150) = 0.5 - P(Z-0.98) = 0.5 - 0.3365 = 0.1653$ or 16.53% that average smoker spends more than $150 per month. $P(Z-0.98)$ can be found in the table of z-score probability values.

It is given that all survey participants have mean age of 28 and standard deviation of 13. This can be checked by taking a sub-sample of 40 participants to test the average age. Assuming that the claim is true, the probability that the sample mean has an average age of 30 or above needs to be calculated. In order to calculate this probability, standard error $(\sigma_x)$ as well as z-value are required as expressed in (6) and (7)

$$\sigma'_x = \frac{\sigma'}{\sqrt{n}} = \frac{30}{\sqrt{40}} = 4.74 \tag{6}$$

$$z = \frac{x - xbar}{\sigma'x} = \frac{30 - 28}{4.74} = 0.42 \tag{7}$$

where $n$ indicates sub-sample of 40 participants, $x$ average age of 30 for participants of the survey, $x_{bar}$ average mean of 28 based on the created dataset. Therefore, $P(x_{bar} >$

*30)* $= 0.5 - P(Z < 0.42) = 0.5 - 0.1628 = 0.3372$ or the probability that selected sample has an average age of 30 or above is 33.72%.

Another goal was to construct a 95% confidence interval for the average monthly spending for both male and female smokers. The easiest way to construct a confidence interval is to use bar charts with the representation of confidence intervals. The confidence interval for male and female smokers is illustrated in Fig. 6.

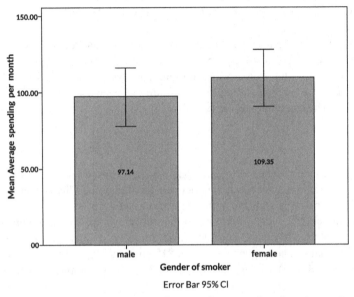

**Fig. 6.** Confidence interval for average spending on tobacco products based on male and female participants of the survey.

With 95% confidence, it can be said that the true mean of average monthly spending on cigarettes for men lies at approximately in the interval between 75–125, whereas for women this interval is approximately between 90–125. Thus, it is evident that women tend to spend more on cigarettes than men on a monthly basis.

## 4   Hypothesis Testing

According to [23] hypothesis testing methodology is widely used to verify the efficacy of an intervention by testing it against the null hypothesis. This has been one of the prominent models used to examine the cause-effect relationship between the use of tobacco and the patient's health and wellbeing outcomes. The first step in hypothesis testing is to clearly state a null and alternative hypothesis. These two hypotheses are tested using a one-tailed or two-tailed test. After computing probability and test statistics, it is necessary to construct a confidence interval based on the chosen testing approach. The final step represents decision making which helps the researcher accept and reject the null/alternative hypothesis. The final decision is made by doing a comparison between the subjective criterion and objective test or created probability statements. As noted in [24] most common research method in hypothesis testing is the Analysis of Variance (ANOVA). This statistical method compares the means of several data samples. It can be considered as an extension to t-test for testing two or more independent samples in the created population. One-way ANOVA is particularly useful when collected data are divided into a specific group based on only one factor.

ANOVA hypothesis testing requires the following assumptions:

- All observations are independent
- All observations in each group come from the normal distribution
- The variance of the population is the same in each group

Furthermore, hypothesis testing method has been prominently used by Leon *et al.* [25] to examine the effect of tobacco in reducing antipsychotic side effects and schizophrenia symptoms. The study examines the reliability of the self-medication hypothesis, which argues that schizophrenic patients smoke to reduce their signs and symptoms. Hence using akathisia to examine its association with heavy smoking, this research sampled 250 *DMS-IV* schizophrenic patients, of which all are Caucasian with the mean age of 36.1 years and are predominantly (78%) males. The analysis of the results indicates that akathisia is not associated with heavy smoking ($p = 0.86$), suggesting the results support the null hypothesis indicating lack of intervention's efficacy. This has been examined against treatments that induce akathisia including the use of typical antipsychotics ($OR = 9.1$; $CI$, 1.2–70.6) and benzodiazepines ($OR = 0.27$, $CI$, 0.08–0.87). In order to examine all aspects of the theory, the authors widened the definition of akathisia, following the Barnes Global score $> 0$. Despite the broader definition, the statistical yield fails to support the self-medication theory ($p = 0.63$). Upon further examination, this study highlights the association between heavy smoking and excited symptoms ($p < 0.001$). As such, those who smoke heavily (61%) exhibited excited symptoms without akathisia compared to those who do not (34%). However, it is worthy to note that heavy smokers with excited symptoms are on lower doses of antipsychotic ($OR = 5.2$; $CI$, 1.9–13.8).

*Hypothesis#1:* One-way ANOVA method

Null hypothesis ($H_0$): *Students spent equal money on cigarettes than others with different profession.*

Alternative hypothesis ($H_a$): *Students spend more money on cigarettes than others with different professions.*

After running a one-way ANOVA test in IBM SPSS statistical software environment, we obtained the following results described in Tables 3 and 4.

From Table 3, we can see that on average students spend less money than any other profession. In Table 4, it can be seen that Sig. a probability value ($p$) equals to 0.521. ($p > 0.05$) is non-significant and we fail to reject the $H_0$ because there is not enough statistical evidence to deny the initial claim of $H_0$. Meaning that there is no difference in spending money on tobacco products in case of students and persons with different professions at a 5% significance level. The $p$ parameter helps in making the decision

**Table 3.** The data distribution of average monthly spending on cigarettes grouped by survey participants of different professions

| | N | Mean | Std. Deviation | Std. Error | 95% Confidence interval for Mean | | Minimum | Maximum | Between-component variance |
|---|---|---|---|---|---|---|---|---|---|
| | | | | | Lower bound | Upper bound | | | |
| Student | 26 | 94.2308 | 33.72513 | 6.61404 | 80.6089 | 107.8526 | 20.00 | 150.00 | |
| Professor | 2 | 150.0000 | 70.71068 | 50.00000 | −485.3102 | 785.3102 | 100.00 | 200.00 | |
| Waiter | 5 | 110.0000 | 54.77226 | 24.49490 | 41.9913 | 178.0087 | 50.00 | 200.00 | |
| Architect | 1 | 100.0000 | . | . | . | . | 100.00 | 100.00 | |
| Engineer | 1 | 150.0000 | . | . | . | . | 150.00 | 150.00 | |
| Others | 17 | 110.5882 | 61.28525 | 14.86386 | 79.0783 | 142.0982 | 10.00 | 200.00 | |
| Total | 52 | 104.4231 | 47.54414 | 6.59319 | 91.1867 | 117.6594 | 10.00 | 200.00 | |
| Model Fixed effects | | | 47.89679 | 6.64209 | 91.0532 | 117.7929 | | | |
| Model Random effects | | | | 6.64209[a] | 87.3490[a] | 121.4971[a] | | | −52.26046 |

**Table 4.** One-way ANOVA test results together with calculated between groups and withing groups parameters for average smoker spending per month

| ANOVA | | | | | |
|---|---|---|---|---|---|
| Average spending per month | | | | | |
| | Sum of squares | df | Mean square | F | Sig. |
| Between groups | 9753.959 | 5 | 1950.792 | 0.850 | 0.521 |
| Within groups | 105528.733 | 46 | 2294.103 | | |
| Total | 1 15282.692 | 51 | | | |

which hypothesis should be accepted or rejected. Also, the alternative hypothesis cannot be accepted because data do not favor the initial claim.

*Hypothesis#2:* Two-tailed *t*-test method

Null hypothesis ($H_0$): *Smokers spend $200 on cigarettes per months.*
Alternative hypothesis ($H_a$): *Smokers do not spend $200 on cigarettes per month.*

A two-tailed statistical test is used in the case when it is necessary to determine if there is any difference between the two observed groups. For instance, here we want to see whether smokers in the created dataset spend exactly $200 on cigarettes per months. The obtained results are listed in Table 5 which suggest that smokers on average spend $104.4 on cigarettes per month ($p < 0.05$). This indicates that the null hypothesis is rejected, whereas the initial claim ($H_a$) is true and can be accepted:

**Table 5.** The results obtained after applying two-tailed *t*-test considering average smoker spending per month

| | N | Mean | Std. Deviation | Std. Error Mean | | | |
|---|---|---|---|---|---|---|---|
| Average spending per month | 52 | 104.4231 | 47.54414 | 6.59319 | | | |

| | | | Test value = O | | | | |
|---|---|---|---|---|---|---|---|
| | t | df | Sig. (2-tailed) | Mean Difference | | 95% Confidence interval of the difference | |
| | | | | | Lower | | Upper |
| Average spending per month | 15.838 | 51 | 0.000 | 104.42308 | 91.1867 | | 117.6594 |

**Table 6.** The data distribution of the average tobacco packets consummation per day grouped by male and female gender

| | | N | Mean | Std. Deviation | Std. Error | 95% Confidence interval for Mean | | Minimum | Maximum | Between-component variance |
| --- | --- | --- | --- | --- | --- | --- | --- | --- | --- | --- |
| | | | | | | Lower bound | Upper bound | | | |
| Male | | 21 | 1.5238 | 0.78224 | 0.1707 | 1.1677 | 1.8799 | 0.50 | 3.00 | |
| Female | | 31 | 1.4516 | 0.90696 | 0.1629 | 1.1189 | 1.7843 | 0.50 | 4.00 | |
| Total | | 52 | 1.4808 | 0.85154 | 0.11809 | 1.2437 | 1.7178 | 0.50 | 4.00 | |
| Model | Fixed effects | | | 0.85925 | 0.11916 | 1.2414 | 1.7201 | | | |
| | Random effects | | | | 0.11916[a] | −0.0333[a] | 2.9948[a] | | | −0.02688 |

*Hypothesis#3:* One-way ANOVA method

Null hypothesis ($H_0$): *Male participants smoke less than female.*

Alternative hypothesis ($H_a$): *Male participants smoke more than female.*

As it can be seen in Table 6 and 7 on average male participants consume 1.52 cigarette packets per day, whereas female participants consume 1.45 packets daily. In one-way ANOVA table Sig. value ($p = 0.767$) is greater than α ($p > 0.05$). therefore, sufficient statistical significance is not available to reject $H_0$. This indicates that there is no difference in average cigarette packs smoking between male and female survey participants even though we assume that males tend to smoke more than females.

**Table 7.** One-way ANOVA test together with calculated between groups and within groups parameters for cigarette packets consummation per day

| ANOVA | | | | | |
|---|---|---|---|---|---|
| Packets consumed daily | | | | | |
| | Sum of squares | df | Mean square | F | Sig. |
| Between groups | 0.065 | 1 | 0.065 | 0.088 | 0.767 |
| Within groups | 36.916 | 50 | 0.738 | | |
| Total | 36.981 | 51 | | | |

*Hypothesis#4:* Correlation assumption + Linear regression

Null hypothesis ($H_0$): *Smokers who consume more packs of a certain cigarette brand are not going to spend more money.*

Alternative hypothesis ($H_a$): *Smokers who consume more packs of a certain cigarette brand are going to spend more money.*

In the correlation matrix, each attribute is adequately correlated with itself, and so parameter $r = 1$ the diagonal of the table. The SPSS marks any correlation coefficient significant at this level with an asterisk. Hence, it can be seen that the attributes "Average spending per month" and "Packets consumed daily" are significantly correlated. Also "Gender of smoker" and "Quitting" are correlated as well, whereas "Profession of the smoker" and "Marital Status" are significantly negatively correlated. The significant correlation between health issues and marital status is also noticeable. As it was determined that attributes "Average spending per month" and "Packets consumed daily" have significant correlation, linear regression needs to be conducted on these attributes by defining the independent variables as packets consumed daily, and dependent as spending per month. In Fig. 7, linear regression between these two variables is illustrated.

**Table 8.** The SPSS matrix indicating generated correlations between attributes in the dataset

| | | Average spending per month | Gender of smoker | Packets consumed daily | Profession of the smoker | Marital status of smoker | Quitting? | Reason to quit? | Favourite cigarette brand | Reason for smoking | Health issues? |
|---|---|---|---|---|---|---|---|---|---|---|---|
| Average spending per month | Pearson correlation | 1 | 0.127 | 0.770 | 0.148 | −0.066 | 0.148 | 0.131 | 0.068 | 0.131 | 0.119 |
| | Sig. (1-tailed) | | 0.184 | 0.000 | 0.147 | 0.321 | 0.148 | 0.177 | 0.315 | 0.177 | 0.201 |
| | N | 52 | 52 | 52 | 52 | 52 | 52 | 52 | 52 | 52 | 52 |
| Gender of smoker | Pearson correlation | 0.127 | 1 | −0.042 | −0.250 | 0.186 | 0.271 | 0.081 | 0.136 | 0.044 | 0.141 |
| | Sig. (1-tailed) | 0.184 | | 0.384 | 0.037 | 0.093 | 0.026 | 0.285 | 0.167 | 0.378 | 0.159 |
| | N | 52 | 52 | 52 | 52 | 52 | 52 | 52 | 52 | 52 | 52 |
| Packets consumed daily | Pearson correlation | 0.770 | −0.042 | 1 | 0.267 | −0.179 | −0.065 | 0.215 | −0.088 | −0.058 | −0.071 |
| | Sig. (1-tailed) | 0.000 | 0.384 | | 0.028 | 0.103 | 0.324 | 0.063 | 0.267 | 0.343 | 0.309 |
| | N | 52 | 52 | 52 | 52 | 52 | 52 | 52 | 52 | 52 | 52 |

(continued)

**Table 8.** (*continued*)

| | | Average spending per month | Gender of smoker | Packets consumed daily | Profession of the smoker | Marital status of smoker | Quitting? | Reason to quit? | Favourite cigarette brand | Reason for smoking | Health issues? |
|---|---|---|---|---|---|---|---|---|---|---|---|
| Profession of the smoker | Pearson correlation | 0.148 | -0.250 | 0.267 | 1 | -0.464 | 0.014 | 0.186 | -0.003 | -0.065 | -0.275 |
| | Sig. (1-tailed) | 0.147 | 0.037 | 0.028 | | 0.000 | 0.461 | 0.094 | 0.491 | 0.324 | 0.024 |
| | N | 52 | 52 | 52 | 52 | 52 | 52 | 52 | 52 | 52 | 52 |
| Marital status of smoker | Pearson correlation | -0.066 | 0.186 | -0.179 | -0.464 | 1 | 0.185 | -0.261 | 0.061 | -0.077 | 0.259 |
| | Sig. (1-tailed) | 0.321 | 0.093 | 0.103 | 0.000 | | 0.095 | 0.031 | 0.333 | 0.294 | 0.032 |
| | N | 52 | 52 | 52 | 52 | 52 | 52 | 52 | 52 | 52 | 52 |
| Quitting? | Pearson correlation | 0.148 | 0.271 | -0.065 | 0.014 | 0.185 | 1 | 0.149 | 0.131 | -0.050 | 0.181 |
| | Sig. (1-tailed) | 0.148 | 0.026 | 0.324 | 0.461 | 0.095 | | 0.146 | 0.177 | 0.364 | 0.100 |
| | N | 52 | 52 | 52 | 52 | 52 | 52 | 52 | 52 | 52 | 52 |
| Reason to quit? | Pearson correlation | 0.131 | 0.081 | 0.215 | 0.186 | -0.261 | 0.149 | 1 | -0.109 | 0.054 | 0.051 |

(*continued*)

**Table 8.** (*continued*)

| | | Average spending per month | Gender of smoker | Packets consumed daily | Profession of the smoker | Marital status of smoker | Quitting? | Reason to quit? | Favourite cigarette brand | Reason for smoking | Health issues? |
|---|---|---|---|---|---|---|---|---|---|---|---|
| | Sig. (1 *tailed) | 0.177 | 0.285 | 0.063 | 0.094 | 0.031 | 0.146 | | 0.221 | 0.353 | 0.360 |
| | N | 52 | 52 | 52 | 52 | 52 | 52 | 52 | 52 | 52 | 52 |
| Favourite cigarette brand | Pearson correlation | 0.068 | 0.136 | −0.088 | −0.003 | 0.061 | 0.131 | −0.109 | 1 | 0.025 | 0.059 |
| | Sig. (1-tailed) | 0.315 | 0.167 | 0.267 | 0.491 | 0.333 | 0.177 | 0.221 | | 0.429 | 0.338 |
| | N | 52 | 52 | 52 | 52 | 52 | 52 | 52 | 52 | 52 | 52 |
| Reason for smoking | Pearson correlation | 0.131 | 0.044 | −0.058 | −0.065 | −0.077 | −0.050 | 0.054 | 0.025 | 1 | −0.062 |
| | Sig. (1-tailed) | 0.177 | 0.378 | 0.343 | 0.324 | 0.294 | 0.364 | 0.353 | 0.429 | | 0.332 |
| | N | 52 | 52 | 52 | 52 | 52 | 52 | 52 | 52 | 52 | 52 |
| Health issues? | Pearson correlation | 0.119 | 0.141 | −0.071 | −0.275 | 0.259 | 0.181 | 0.051 | 0.059 | −0.062 | 1 |
| | Sig. (1-tailed) | 0.201 | 0.159 | 0.309 | 0.024 | 0.032 | 0.100 | 0.360 | 0.338 | 0.332 | |
| | N | 52 | 52 | 52 | 52 | 52 | 52 | 52 | 52 | 52 | 52 |

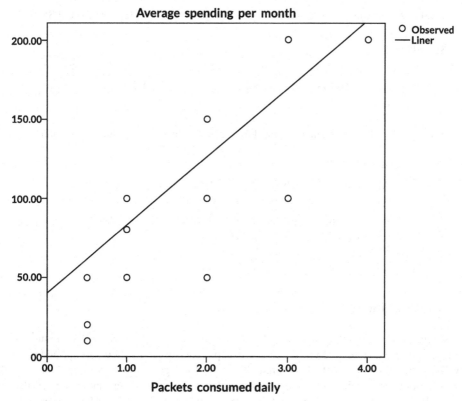

**Fig. 7.** Linear regression between average cigarette spending per month and daily packet consumption attributes in the created dataset

In Table 9 the value of $R = 0.770$ indicates the simple correlation between packets consumed and average monthly spending, which was confirmed by correlation results as described in Table 8. The value of $R_2 = 0.584$ indicates that cigarette packets consumption can account for 58.4% of the variation of monthly spendings. This means that nearly 42% of the variation in monthly spendings are due to the other factors and they cannot be described by only cigarette packets consumption. Other variables from the dataset have their influence as well. In Table 10, it is noticeable that $F$ is significant at $p < 0.001$. This result indicates that there is less than 15% probability that such a large value of $F$-ratio would happen if the null hypothesis is true. Therefore, we can see that the linear regression model results in significantly better prediction of average monthly spendings compared to the case of using the mean value of monthly spendings. To sum up, the regression model overall predicts average monthly spendings significantly well and stated alternative hypothesis is accepted.

**Table 9.** The summary of the correlation model parameters developed in IBM SPSS

| Model summary | | | | |
|---|---|---|---|---|
| Model | R | R-Square | Adjusted R-Square | Std. Error of the estimate |
| 1 | 0.770[a] | 0.593 | 0.584 | 30.64955 |

**Table 10.** The Regression model created for the prediction of average monthly spending

| ANOVA | | | | | | |
|---|---|---|---|---|---|---|
| Model | | Sum of squares | df | Mean square | F | Sig. |
| 1 | Regression | 68312.958 | 1 | 68312.958 | 72.720 | 0.000[b] |
| | Residual | 46969.735 | 50 | 939.395 | | |
| | Total | 115282.692 | 51 | | | |

## 5  Conclusion

In summary, this research represents the statistical analysis of collected data about the smoking of tobacco between different age groups. The survey method was applied with twelve attributes of interests, which may be useful for organizations to control tobacco consumption on a daily or monthly basis. After processing of unwanted data in the created dataset, some useful probabilities are calculated using Bayes' Theorem including different probability distributions such as Binomial, Hypergeometric and Poisson. Finally, four hypotheses were tested applying ANOVA test, two-tailed t-test and linear regression model based on correlation assumption. The $p$ parameter of ANOVA testing reveals that there is no significant difference of tobacco consumption between students and other people with different including male and female participants. Moreover, the two tails test indicates that smokers do not spend $200 on cigarettes per month. The correlation matrix makes a good prediction of average monthly significant spending and finds that smokers who consume more packs of certain cigarette brand are going to spend more money. Organizations that control smoking and the tobacco industry may apply this statistic knowledge in a specific area or country to improve the limitation of this research's result as this dataset was not large enough.

**Conflict of Interest Declaration..** The authors declare no conflict of interest.

## References

1. Martin, T.: Smoking Statistics From Around the World. Very well Mind. https://www.verywe llmind.com/global-smoking-statistics-2824393. Accessed
2. Ng, M., Freeman, M.K., Fleming, T.D., et al.: Smoking prevalence and cigarette consumption in 187 countries, 1980–2012. JAMA **311**, 183–92 (2014)

3. Zhou, S.L.G.X., et al.: Smoke: A Global History of Smoking. Reaktion Books, Clerkenwell (2004)
4. Hall, W.: Cigarette century: the rise, fall and deadly persistence of the product that defined america (in eng). Tob Control **16**(5), 360–360 (2007). https://doi.org/10.1136/tc.2007.021311
5. Gately, I.: Tobacco: A Cultural History of How an Exotic Plant Seduced Civilization. Grove Atlantic, New York (2007)
6. A. I. o. Health and Welfare: "Australia's health 2018." AIHW, Canberra (2018). https://www.aihw.gov.au/reports/australias-health/australias-health-2018
7. Bergen, A.W., Caporaso, N.: Cigarette smoking. JNCI J. Nat. Cancer Inst. **91**(16), 1365–1375 (1999). https://doi.org/10.1093/jnci/91.16.1365
8. West, R.: Tobacco smoking: health impact, prevalence, correlates and interventions (in eng). Psychol. Health **32**(8), 1018–1036 (2017). https://doi.org/10.1080/08870446.2017.1325890
9. Ribeiro Sarmento, D., Yehadji, D.: An analysis of global youth tobacco survey for developing a comprehensive national smoking policy in Timor-Leste. BMC Publ. Health **16**(1), 65 (2016). https://doi.org/10.1186/s12889-016-2742-5
10. Nassar, H.: The economics of tobacco in egypt: a new analysis of demand. Center for Tobacco Control Research and Education. University of California at San Francisco, UC San Francisco (2003)
11. World Health Organization: WHO Report on the Global Tobacco Epidemic, 2009: Implementing Smoke-free Environments. World Health Organization, Geneva (2009)
12. World Health Organization: Brief profile on tobacco health warnings in the South-East Asia Region (2009)
13. U. C. f. D. C. a. Prevention. Youth and Tobacco Use. https://www.cdc.gov/tobacco/data_statistics/fact_sheets/youth_data/tobacco_use/index.htm. Accessed
14. Corey, C.G., Ambrose, B.K., Apelberg, B.J., King, B.A.: Flavored tobacco product use among middle and high school students—United States, 2014. Morb. Mortal. Wkly Rep. **64**(38), 1066–1070 (2015)
15. Paraje, G., Araya, D., Drope, J.: The association between flavor capsule cigarette use and sociodemographic variables: evidence from chile (in eng). PLoS ONE **14**(10), e0224217–e0224217 (2019). https://doi.org/10.1371/journal.pone.0224217
16. World Health Organization: 10 facts on gender and tobacco. https://www.who.int/gender/documents/10facts_gender_tobacco_en.pdf. Accessed
17. Giovino, G.A., et al.: Tobacco use in 3 billion individuals from 16 countries: an analysis of nationally representative cross-sectional household surveys. Lancet **380**(9842), 668–679 (2012). https://doi.org/10.1016/S0140-6736(12)61085-X
18. Piñeiro, B., et al.: Gender differences in use and expectancies of e-cigarettes: online survey results (in eng). Addict. Behav. **52**, 91–97 (2016). https://doi.org/10.1016/j.addbeh.2015.09.006
19. Chen, L.-S., Kaphingst, K., Tseng, T.-S., Zhao, S.: How are lung cancer risk perceptions and cigarette smoking related?-Testing an accuracy hypothesis. Translational Cancer Research (2016)
20. Zhang, Y., Liu, J., Zhang, Z., Huang, J.: Prediction of daily smoking behavior based on decision tree machine learning algorithm. In: 2019 IEEE 9th International Conference on Electronics Information and Emergency Communication (ICEIEC), Beijing, China, pp. 330–333 (2019)
21. Check, J., Schutt, R.K.: Survey research. In: Check, J., Schutt, R.K., (eds) Research Methods in Education. Sage Publications, Thousand Oaks, pp. 159–185 (2012)
22. Hayes, A.: "Bayes' Theorem Definition" (2019). https://www.investopedia.com/terms/b/bayes-theorem.asp
23. Emmert-Streib, F., Dehmer, M.: Understanding statistical hypothesis testing: the logic of statistical inference. Mach. Learn. Knowl. Extr. **1**(3), 945–961 (2019)

24. Ostertagova, E., Ostertag, O.: Methodology and application of one-way ANOVA. Am. J. Mech. Eng. **1**, 256–261 (2013)
25. de Leon, J., Diaz, F.J., Aguilar, M.C., Jurado, D., Gurpegui, M.: Does smoking reduce akathisia? Testing a narrow version of the self-medication hypothesis. Schizophr. Res. **86**(1), 256–268 (2006). https://doi.org/10.1016/j.schres.2006.05.009

# The Influence of Stringency Measures and Socio-Economic Data on COVID-19 Outcomes

Azra Musić⬤, Jasminka Hasić Telalović$^{(\boxtimes)}$ ⬤, and Dženita Đulović⬤

University Sarajevo School of Science and Technology, Hrasnicka Cesta 3a, Sarajevo,
Bosnia and Herzegovina
jasminka.hasic@ssst.edu.ba

**Abstract.** The COVID-19 pandemic has changed the way that society functions. The state governments have chosen different measures to respond to the spread of the virus as there was no clear evidence on what the best way is to respond. In this paper, we gather the evidence of governments' stringency measures and couple it with countries' socio-economic data (including the amounting of tests that each country performed) to explore how they influenced the COVID-19 spread. We evaluated how well seven different regression models can predict the spread of the virus. The spread of the virus is expressed through the number of positively identified individuals as well as deaths reported due to the virus. For the prediction of the number of new cases, ElasticNet algorithm was the best performing, followed by Linear Regression and LASSO. LASSO algorithms predicted the best number of new deaths. The metrics used to evaluate were: R-squared score, Adjusted R-squared Score, mean square error (MSE), mean absolute error (MAE), and root mean square error (RMSE). The most effective measure was the stay-at-home measure followed by workplace and school closures. With the availability of new and more accurate data and variables, the models can be further improved.

**Keywords:** COVID-19 · Government response · Regression · Stringency measures

## 1 Introduction

The COVID-19 pandemic has a deep impact on societies. The governments are faced with making the decisions to best address the circumstances for their constituents. As this is a novel virus and no similar pandemic has happened in modern times, there is a very little experience that governments can rely on when making stringency conditions. This has resulted in a great range of stringency responses. Some countries imposed strict lockdowns while others imposed very few measures besides keeping social distance. The more severe the stringency measures are, the more under the control the spread of the virus is but also, the disruption on the society functions. As the threat of pandemic has been raging throughout the world for over six months now, there is also substantial evidence on how different government stringency measures have affected the spread of

© Springer Nature Switzerland AG 2021
J. Hasic Telalovic and M. Kantardzic (Eds.): MeFDATA 2020, CCIS 1343, pp. 39–54, 2021.
https://doi.org/10.1007/978-3-030-72805-2_3

the virus. In the literature review, while we found a number of studied that propose models that forecast the spread of the virus, we did not find an analysis that uses government stringency measures into account.

Our aim in this study was to collect and analyze the data that can be used to forecast the future spread of the virus, and as such, aid the governments in the decision making for the control of the virus spread.

In this study, we collected the data from different current sources and constructed a dataset that can be used to effectively study COVID-19 spread in different countries. We use both well established as well as novel data sources. For the independent variables, we started with socio-economic data, added the number of COVID-19 tests that were performed, and finally augmented the dataset with stringency measures. Our dependent variables are measures of the COVID-19 virus spread (which include the number of cases as well as fatalities reported to be caused by this virus).

## 2   Background

### 2.1   Related Work

Several papers have applied various data science techniques to analyze and predict COVID-19 parameters and outcomes. Ellen [3] and Latif et al. [14] wrote review papers on the usage of mathematical modeling and data science to address the spread and impact of the COVID-19 virus.

In [3] the following lessons learned are outlined:

- With no measures taken, the virus spreads exponentially,
- Without immunization, COVID-19 will be around for a long time,
- Mobility control is a drastic but effective measure,
- There is a two-week delay between mobility and reproduction,
- Massive amounts of data are available, but they do not always align in the models,
- Selective opening measures can be more effective than voluntary quarantine,
- For safe reopening, testing is critical.

[14] adds the following items to the list of challenges:

- Limitation of data,
- Urgency vs. correctness of results
- Security, privacy, and ethics,
- The necessity of multidisciplinary collaborations,
- Novel data modalities,
- The need for solutions for the developing world.

Besides, [3] summarized the following studied topics: Risk Assessment and Patient Prioritization, Screening and Diagnosis, Simulation and Modelling, Contact Tracing, Understanding Social Interventions, Logistical Planning and Economic Interventions, Automated Patient Care, Supporting Vaccine Discovery and New Treatments.

Ahmed ben et al. [1] used deep learning to predict the number of new daily cases. Their model has been applied to the country of Qatar. Their findings show a significant increase in cases if school operations are not restricted and borders are open.

The R0 mathematical model (Next-generation method) was applied in Kamran and Ghader [12] to model the spread of the virus. Also, the Runge-Kutta method was used for solving the governing nonlinear equations. The models have been tested on data from Iran and Italy. The aim of Anna et al. [2] was to combine open-source data from different sources to discover changes in mobility in Europe. The findings show that citizens drove more and walked less.

Several ML techniques (linear regression – LR, LASSO, support vector Ma-chines – SVM, and exponential smoothing – ES) were utilized in Rustam [17] to predict the number of new cases, deaths, and recoveries in the next 10 days. An already existing dataset was utilized. Different ML algorithms had the best performance on different datasets. ES performed best when the time-series dataset has very limited entries.

The disruption of economic processes is a big challenge caused by this pandemic. Nikolopoulos K. et al. [15] analyzed the disruptions of the supply chain. They used data from the USA, India, UK, Germany, and Singapore up to mid-April 2020. The data on-demand quantities were gathered from Google trends of four different sectors (Groceries, Electronics, Fashion, Automotive). Clustering and Partial Curves and Nearest Neighbor Forecasting (CPC–NN) was used for forecasting. The results show that the change in demand depends on the timing of the lockdown and type of product. The earlier a lockdown is imposed, the higher the excess demand will be for groceries, and the longer the lockdown lasts the higher the cumulative excess demand, and thus the higher the need for planning for production and inventory.

The analysis of people flow using the extension theory to predict and prevent risk was presented in He [8]. Flow has been graded into four levels: below normal, general, larger, and large. Three situations were analyzed: severe regional epidemic, local epidemic, and no epidemic.

Yves et al. [22] used the data from the Center for Systems Science and Engineer-ing (CSSE) at Johns Hopkins University and air transport (passengers carried) from the World Bank to study the influence of air traffic on the spread of COVID-19. The follow-ing models were utilized: Poisson model (PM), Quasi-Poisson model (QPM), Negative binomial model (NBM), zero-inflated models (ZIM), and Hurdle models (HM). A direct link was established between air traffic increases and the number of new infections.

The dataset analyzed in Sina et al. [18] is based on data collected in Italy, Germany, Iran, the USA, and China. It was used to predict the outbreak of the COVID-19 virus. The evaluation was conducted using the root mean square error (RMSE) and correlation coefficient, and two models showed promising results (i.e., multi-layered perceptron, MLP, and adaptive network-based fuzzy inference system, ANFIS).

## 2.2 Regression Algorithms

Machine learning techniques include regression algorithms. Regression algorithms examine a number of input (independent) and output (dependent) variables and model the relationship between them. For the establishment of the model, different mathemat-ical fittings are applied. The result is a function that expresses the dependence of input and output variables which can take as an input a new set of input variables and predict the outputs.

In this study, we examined seven algorithms: Linear regression, Logistic regression, Polynomial regression, Stepwise regression, Ridge regression, and Lasso regression. More details about these algorithms can be found in [6, 9, 13].

## 3 Dataset

The dataset that was constructed for this study was assembled using both established sources for important country indicators as well as novel sources that have bud since the COVID-19 pandemic started. The list of variables that were initially collected to be analyzed for this study is summarized in Table 1.

We first summarize the novel sources. In [11] the stringency response of governments was collected in each was categorized into four levels of stringency. The 17 indicators include containment and closure policies, economic policies, and health system policies. Worldmeter [24] is a highly relevant source and trusted authority for COVID-19 statistics for countries worldwide. The historical data from this website was obtained through web archives. The European Center for Disease and Prevention and Control [4] publishes various guidelines, research papers, and is constantly keeping track of the data regarding the spread of COVID-19 in the EU as well as worldwide.

To characterize countries, we used several variables also from the established sources. Those include International Standardization Organization [10], The World of Data [7], United Nations data [20, 21], World Bank data [23], GBD 2017 [5], and for life-expectancy estimates we used [11].

**Table 1.** Dataset variables, their sources, and types.

| Variable description | Source | Variable type |
|---|---|---|
| Confirmed cases per 1 M people | [4] | Dependent |
| New confirmed cases per 1 M people | [4] | Dependent |
| Total deaths per 1 M people | [4] | Dependent |
| New deaths per 1 M people | [4] | Dependent |
| Government Response Stringency Index | [19] | Independent |
| School-closures stringency | [19] | Independent |
| Workplace-closures stringency | [19] | Independent |
| Public-events stringency | [19] | Independent |
| Public-gathering-rules stringency | [19] | Independent |
| Public-campaigns stringency | [19] | Independent |
| Stay-at-home (lockdowns) restriction | [19] | Independent |
| Public transport stringency | [19] | Independent |

(*continued*)

**Table 1.** (*continued*)

| Variable description | Source | Variable type |
|---|---|---|
| Internal-movement stringency | [19] | Independent |
| International travel | [19] | Independent |
| Testing-policy | [19] | Independent |
| Contact-tracing | [19] | Independent |
| Life expectancy at birth in 2019 | [11] | Independent |
| Number of COVID-19 tests per capita | [24] | Independent |
| ISO country codes | [10] | Independent |
| Continent of the geographical location | [7] | Independent |
| Date of observation | [7, 24] | Independent |
| Population in 2020 | [21] | Independent |
| Population density | [23] | Independent |
| GDP | [23] | Independent |
| Population % living in extreme poverty | [23] | Independent |
| Number of hospital beds per 1 K people | [23] | Independent |
| Number of air passengers per capita | [23] | Independent |
| Median age of the population | [20] | Independent |
| Population that is 65 years and older | [20] | Independent |
| Death rate from cardiovascular disease in 2017 | [5] | Independent |

## 4  Methods

The aim of this study is to future forecast the number of confirmed COVID-19 infected cases and the number of deaths caused by the COVID-19 daily for the European countries. The additional aim was to extract the measures that were highly correlated with the changes in the number of confirmed cases and deaths.

The forecasting was done using seven supervised machine learning approaches, considered to be appropriate for this type of analysis (Linear regression, Logistic regression, Polynomial regression, Stepwise regression, Ridge regression, and Lasso regression). The constructed dataset contains daily time series that include COVID-19 related statistical data, sociodemographic characteristics of countries, and data on governmental response measures. Initially, the data for European countries has been extracted as that region is characterized by similar governmental approaches due to the European Union, and similar political orientations of the neighboring countries. Among countries extracted, the countries with a population of higher than 1 million have been selected for further processing. Moreover, variables that had more than 20% of missing values for the given state were excluded, so the total number of features used in the initial correlation analysis was 14. Those included: total cases per million, new cases per million, total deaths per million, new deaths per million, tests per million, stringency index, population density, median age, population older than 65, GDP per capita, percent of the population living in extreme poverty, cardiovascular death rate, hospital beds per thousand, and life expectancy. The sociodemographic variables that had a moderate or high positive or negative correlation with COVID-19 data were used for the model development, along with the governmental response metrics. A moderate correlation is between values ±0.3 to ±0.7, and a high correlation is for values above/below ±0.7 (Fig. 1).

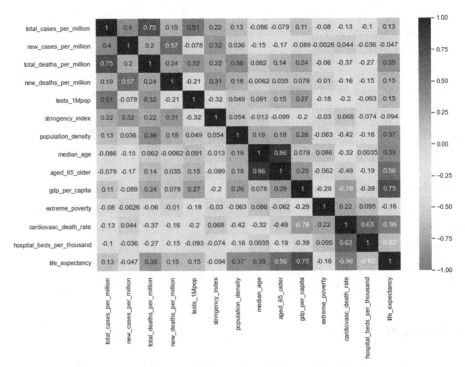

**Fig. 1.** Heatmap plot representing the correlation matrix between numerical variables to be considered for the data model. The positive values indicate positive relationship between two variables, while negative indicate negative relationship between variables.

The extracted variables include dependent ones – total and new cases per million, total and new deaths per million, and independent – tests per million, stringency index, population density, cardiovascular death rate, and life expectancy. The variables were scaled using MinMaxScaler and added to the preprocessed dataset response. Those variables were encoded using one-hot encoding to better capture the categorization of different responses. The response measures variables finally added to the dataset ready for the modeling were: testing policy, contact tracing, internal movement, public campaigns, public events, public gathering, public transport, school closures, stay-at-home, workplace closures, and international movement. The changes in levels of measures applied over time for top five countries with the highest number of cases and deaths (Spain, France, Great Britain, Germany, Italy) and countries of Western Balkans (excluded

Montenegro and Northern Macedonia due to the lack of data) are demonstrated in the Appendix.

After the initial data preprocessing step, the dataset was divided into two subsets used for training (80%) and testing (20%). The models – linear regression, logistic regression, polynomial regression, stepwise regression, ridge regression, LASSO regression, and ElasticNet regression, were trained on the number of cases, tests, and deaths patterns. The performance evaluation was done in terms of important measures including R-squared score (R2 score), Adjusted R-squared Score (R2 adjusted), mean square error (MSE), mean absolute error (MAE), and root mean square error (RMSE). The proposed pipeline used in the study has been shown in Fig. 2 as a block diagram.

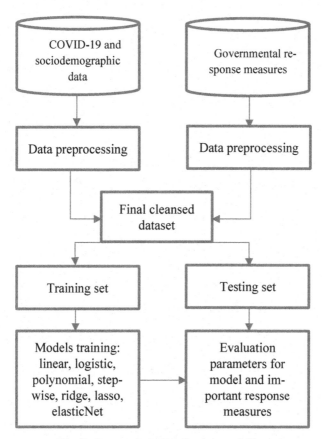

**Fig. 2.** Proposed pipeline for data modelling

## 5   Results

This study aims to develop a pipeline for forecasting the number of COVID-19 cases and deaths, by using sociodemographic data and governmental response measures data as input. In addition, it underlines the most important response measures depending on their weight in generating regression function.

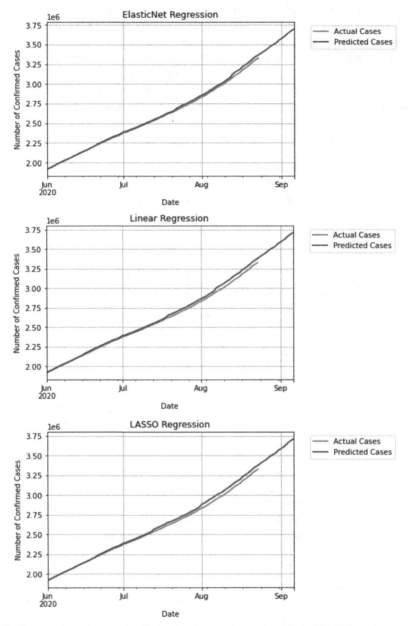

**Fig. 3.** The number of cases prediction by the top three algorithms (ElasticNet. Linearn and LASSO Regression)

The first prediction evaluated is on the number of cases, and according to the results *ElasticNet* performed the best among all the models. *Linear Regression* and *LASSO* performed equally well, while the worst performance had the *Logistic* and *Stepwise Regression* models. The results are shown in Table 2.

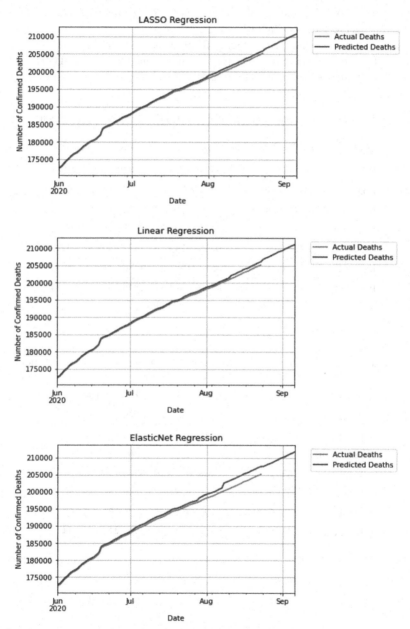

**Fig. 4.** The number of deaths prediction by the top three algorithms (LASSO. Linear and ElasticNet Regression)

Figure 3 shows the performance of the three best performing models – *ElasticNet, Linear,* and *LASSO* respectively. Graphs predict that the number of cases will slightly increase in the upcoming two weeks' time period.

We then evaluated the prediction of the number of deaths, and *LASSO* performed the best among all the models. *Linear Regression* performed almost equally well, while the worst performance had the *Logistic and Stepwise Regression* models in this case as well. The results are shown in Table 3.

Figure 4 summarizes the performance of the three best performing models – *LASSO, Linear,* and *ElasticNet* respectively. The prediction was that the number of deaths would not significantly increase in the next two weeks' time.

**Table 2.** Evaluation parameters for number of COVID-19 positive cases prediction

| Model | $R^2$ | $R^2$ adjusted | MSE | MAE | RMSE |
|---|---|---|---|---|---|
| Linear | 0.81 | 0.78 | 1532344244.21 | 32738.34 | 39145.17 |
| Logistic | 0.42 | 0.30 | 7039509902.03 | 80234.54 | 83901.79 |
| Polynomial | 0.74 | 0.68 | 2847283234.23 | 40285.53 | 53359.94 |
| Stepwise | 0.44 | 0.33 | 6847284294.34 | 78538.35 | 82748.32 |
| Ridge | 0.72 | 0.65 | 3043849249.35 | 45729.60 | 55171.09 |
| Lasso | 0.81 | 0.78 | 1458493585.34 | 31802.43 | 38190.23 |
| Elasticnet | 0.84 | 0.82 | 1043583054.42 | 28485.32 | 32304.54 |

**Table 3.** Evaluation parameters for number of COVID-19 caused deaths prediction

| Model | $R^2$ | $R^2$ adjusted | MSE | MAE | RMSE |
|---|---|---|---|---|---|
| Linear | 0.81 | 0.77 | 2849392.24 | 1049.48 | 1688.01 |
| Logistic | 0.42 | 0.32 | 31843993.84 | 3919.84 | 5643.05 |
| Polynomial | 0.62 | 0.53 | 8384858.11 | 2894.41 | 2895.66 |
| Stepwise | 0.43 | 0.31 | 29849291.10 | 3749.31 | 1727.51 |
| Ridge | 0.67 | 0.58 | 8048284.98 | 2294.43 | 2836.99 |
| Lasso | 0.82 | 0.79 | 2324526.47 | 940.32 | 1524.64 |
| Elasticnet | 0.75 | 0.67 | 5403039.39 | 1895.42 | 2324.44 |

# 6   Discussion

Out of seven tested regression models, *Linear Regression, ElasticNet Regression,* and *LASSO Regression* were the top three model performers in the number of new confirmed cases and deaths forecasting. To extract the most influential response measures, regression equations for those three models have been evaluated and analyzed.

The most influential response measure is the stay-at-home measure, as expected since it includes almost all other policies defined. However, one should be aware that stay-at-home measure should be the ultimate measure, as it highly affects day-to-day lives. With this measure enforced over a long period, the COVID-19 pandemic could be slowed down, but it might lead to economic and psychological trauma to the country's citizens.

The second most influential response measure is workplace closure. Close to this measure is school closure as well since those two are relatively carried out in the near time. Schools shifted to the online lectures, while on-site workplaces were replaced by remote home-offices.

The response measure following workplace closure and school closure is an international movement ban. Since the virus is originally brought to Europe, and the rest of the world, from the Chinese province of Wuhan via international travelers, closing down the border crossings had a high impact of limiting the spread within the countries. It is one of the most recommended measures, to avoid smaller nations having to deal with high peaks of cases they cannot properly handle, as well as to limit any other peaks potentially caused by virus carriers with no to small symptoms present.

Each country has the responsibility to make its response measures by observing the current numbers of cases handled locally. However, one should observe the situation in other, neighbor, and similar-system countries, to learn from their experience and avoid potential mistreatment of political power during these critical times.

# 7   Conclusions

This forecast study can be of great help for the government authorities to take timely actions and make decisions to better respond to the COVID-19 ongoing pandemic. Different stringency measures can be put into the model and the predicted spread of the virus can be evaluated in order not to exceed the medical capacities that the government manages. The forecasts of this study can be further enhanced with the introduction of new data into the dataset and inclusion of wider range temporal data. Furthermore, additional potentially more accurate and appropriate machine learning methods for forecasting could be evaluated.

# Appendix

The figures fellow visualize changes in levels of measures applied over time for top five countries with the highest number of cases and deaths (Spain- ESP, France- FRA, Great Britain - GBR, Germany DEU, Italy - ITA) and countries of Western Balkans (Albania – ALB, Bosnia and Herzerovina – BiH, Croatia – HRV, Serbia - SRB) (Figs. 5–15).

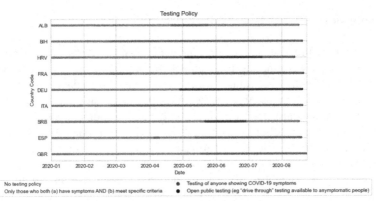

**Fig. 5.** A.1 Testing policy

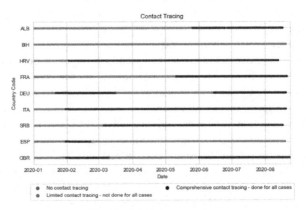

**Fig. 6.** A.2 Contact tracing

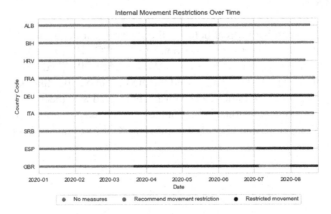

**Fig. 7.** A.3 Internal movement

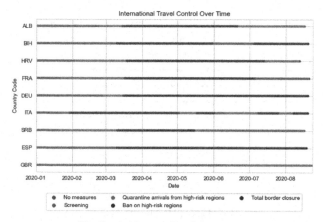

**Fig. 8.** A.4 International movement

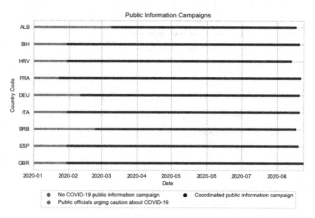

**Fig. 9.** A.5 Public campaigns

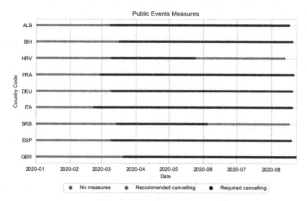

**Fig. 10.** A.6 Public events

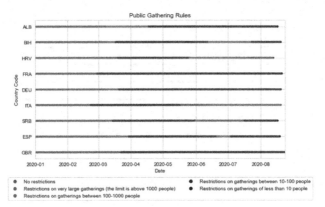

**Fig. 11.** A.7 Public athering

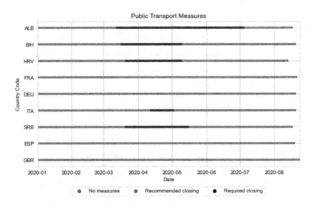

**Fig. 12.** A.8 Public transport

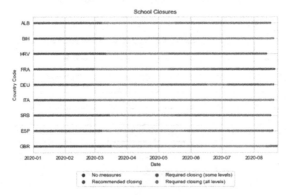

**Fig. 13.** A.9 School closures

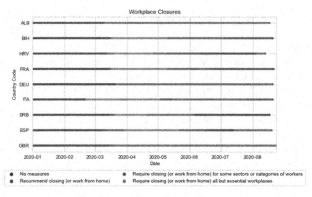

**Fig. 14.** A.10 Workplace closures

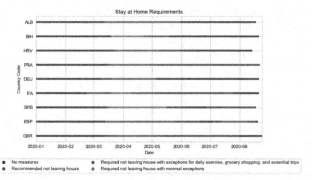

**Fig. 15.** A.11 Stay-at-home measure

# References

1. Ahmed ben, S., Abdelkarim, E., Hussein, A., Abdelmonem, M.: A deep-learning model for evaluating and predicting the impact of lockdown policies on COVID-19 cases. arXiv:2009. 05481v1 (2020)
2. Islind, A.S., Mar√ ≠ a, √ì., Harpa, S.: Changes in mobility patterns in Europe during the COVID-19 pandemic: Novel insights using open source data. arXiv:2008.10505 (2020)
3. Ellen, K.: Data-driven modeling of COVID-19-Lessons learned. Extreme Mech. Lett. **40**, 100921 (2020). ISSN 2352-4316
4. European Centre for Disease Prevention and Control. https://www.ecdc.europa.eu/en/covid-19-pandemic. Accessed 29 Sep 2020
5. GBD 2017 Risk Factor Collaborators: Global, regional, and national comparative risk assessment of 84 behavioural, environmental and occupational, and metabolic risks or clusters of risks for 195 countries and territories, 1990–2017: a systematic analysis for the Global Burden of Disease Study 2017. Lancet **392**, 1923–1994 (2018). https://doi.org/10.1016/S0140-673 6(18)32225-6
6. Gutierrez, D.D.: Machine Learning and Data Science: An Introduction to Statistical Learning Methods with R. Technics Publications, New Jersey (2015)
7. Hannah, R., et al.: Coronavirus Pandemic (COVID-19). Our World in Data (2020). (https:// ourworldindata.org/coronavirus)

8.  He, P.: Study on epidemic prevention and control strategy of COVID -19 based on personnel flow prediction. In: 2020 International Conference on Urban Engineering and Management Science (ICUEMS), pp. 688–691. Zhuhai, China (2020)

9.  Hsieh, W.W.: Machine Learning Methods in the Environmental Sciences: Neural Networks and Kernels. Cambridge University Press, Cambridge (2009)

10. International Standardization Organization [ISO]: ISO 3166-1: 2020(en) Codes for the representation of names of countries and their subdivisions—Part 1: Country code (2020)

11. James, C.R.: Estimates of regional and global life expectancy, 1800–2001. Popul. Dev. Rev. **31**(3), 537–543 (2005)

12. Kamran, S., Ghader, R.: A New Dynamic Model to Predict the Effects of Governmental Decisions on the Progress of the CoViD-19 Epidemic. arXiv:2008.11716 (2020)

13. Kutner, M.H., Nachtsheim, C.J., Neter, J., Li, W.: Applied Linear Statistical Models. McGraw-Hill Irwin, New York (2005)

14. Latif, S., Usman, M., Manzoor, S., Iqbal, W., Qadir, J., Tyson, G., et al.: Leveraging data science to combat COVID-19: a comprehensive review. TechRxiv. Preprint (2020)

15. Nikolopoulos, K., et al.: Forecasting and planning during a pandemic: COVID-19 growth rates, supply chain disruptions, and governmental decisions. European Journal of Operational Research (2020). 10.1016/j.ejor.2020.08.001

16. Paiva, H.M., Afonso, R.J.M., de Oliveira, I.L., Garcia, G.F.: A data-driven model to describe and forecast the dynamics of COVID-19 transmission. Plos One **15**(7), e0236386 (2020)

17. Rustam, F., et al.: COVID-19 future forecasting using supervised machine learning models. IEEE Access **8**, 101489–101499 (2020)

18. Sina, F., et al.: COVID-19 outbreak prediction with machine learning. medRxiv 2020.04.17.20070094 (2020)

19. Thomas, H., et al.: Variation in government responses to COVID-19. Oxford COVID-19 Government Response Tracker, Blavatnik School of Government (2020)

20. United Nations, Department of Economic and Social Affairs, Population Division: World Population Prospects: The 2017 Revision. https://population.un.org/wpp/Publications/Files/WPP2017_DataBooklet.pdf. Accessed 28 Sep 2020

21. United Nations, Department of Economic and Social Affairs, Population Division: World Population Prospects: The 2019 Revision. https://population.un.org/wpp/. Accessed 28 Sep 2020

22. Yves, M.S., Mintod√/™, N.A.: The influence of passenger air traffic on the spread of COVID-19 in the world. Transp. Res. Interdisc. Perspect. **8**, 100213 (2020). ISSN 2590-1982

23. World Bank. https://data.worldbank.org/. Accessed 29 Sep 2020

24. Worldometer.    http://web.archive.org/web/*/https://www.worldometers.info/coronavirus/. Accessed 29 Sep 2020

# Applications in Medicine

# 3D Vessel Segmentation in CT for Augmented and Virtual Reality

Agnese Simoni[1], Eleonora Tiribilli[1,2], Cosimo Lorenzetto[2], Leonardo Manetti[2], Ernesto Iadanza[2] (ID), and Leonardo Bocchi[1(✉)] (ID)

[1] Department of Information Engineering, University of Florence, Via S. Marta 3, 50139 Florence, Italy
leonardo.bocchi@unifi.it
[2] Epica Imaginalis, Via Rodolfo Morandi 13/15, 50019 Sesto Fiorentino, Italy

**Abstract.** Blood vessels 3D rendering has numerous applications, ranging from diagnosis to pre-procedural and surgical approaches as it enriches vessels visualization for the clinician. In this paper, we propose a 3D blood vessels segmentation method designed for use with a cone beam computed tomography (CBCT). The algorithm constitutes a module in the development of an angio-CT visualization system, based on augmented or virtual reality as instruments supporting and improving medical decisions.

The proposed segmentation tool exploits a bone segmentation step for easing the extraction of blood vessels. Both steps are based on region growing technique. For each subject are used two CBCT acquisitions, where the first one is acquired with a traditional CT scan and is used for bone extraction, while the second one is acquired after contrast medium administration and is used for vessels reconstruction.

This novel segmentation algorithm provides an automatic and accurate tool to segment and render blood vessels tree. 3D anatomical models were viewed through a virtual reality environment in order to validate the visualization system.

**Keywords:** Cone bean computed radiography · Tissue segmentation · Virtual and Augmented Reality

## 1 Introduction

Segmentation of anatomical structures is a high interest topic in medical imaging, and numerous applications in computer aided diagnosis (CAD), taking advantages of this technology, have been developed in the past years. In particular, blood vessel analysis and segmentation became crucial not only for diagnosis, but also for treatment, planning, execution and evaluation of clinical outcomes in several fields of medicine, such as oncology [5], ophthalmology [4], neurosurgery [6], as well as in min-invasive surgery techniques such as laparoscopy, endoscopy, robotic and endovascular surgery [10].

J. Hasic Telalovic and M. Kantardzic (Eds.): MeFDATA 2020, CCIS 1343, pp. 57–68, 2021.
https://doi.org/10.1007/978-3-030-72805-2_4

The importance of blood vessel analysis is supported by the constant introduction in clinical practice of new medical technologies aimed at enhancing vessels visualization by the clinician. In this framework, Virtual and Augmented Reality has experienced a steady growth in medicine in recent years [1,7,16]. At the same time, radiological images acquired with contrast media play a central role in the diagnosis and planification of surgical approaches and are the gold standard in the field.

The aim of the study is to segment vessels obtained by a CT acquired after the injection of contrast media, in order to visualize these anatomical structures with a virtual reality system. This tool is designed to be used with a commercial CBCT (Cone Beam Computed Tomography) developed from the company Epica Imaginalis, together with a virtual reality viewer.

As seen above, segmentation techniques have a vast field of application, both in research and in clinical field; this has led to the development of a large number of algorithms and approaches to the problem, as show by the huge amount of papers annually published in this field [13]. The proposed algorithms range from simple and intuitive methods such as threshold, to more complex and specific approaches such as artificial intelligence and machine learning [15]. Tetteh et al. [17] present DeepVesselNet, an architecture tailored to the challenges of extracting vessel networks and network features from MRA dataset using deep learning. Fu et al. [9] formulate the vessel segmentation as a boundary detection problem, and utilize the fully convolutional neural networks (CNNs) to generate a 2D vessel probability map.

Region growing is a versatile and deep-rooted procedure that has been used for several tasks from functional analysis of the left ventricle in multislice CT (MSCT) of the heart [14], to segmentation of single neurons in their own arrangement within the brain [3]. The fundamental idea of region growing is grouping of pixels into increasingly larger regions based on a predefined inclusion criteria. This inclusion criteria can be varied and adapted to the type of data and the problem in question, yielding a very flexible and powerful method.

In contrast to conventional CT, the CBCT system employs a wide-angle cone-beam X-ray source and a large area detector; thus, those are much more affected by X-ray scattering, that is one of the main factors causing a drop in image quality. In addition, longer acquisition time of each rotation leads to temporal inhomogenity of contrast media in the vessel. These two factors are the main cause of the decreased uniformity of vessel (or, in general, tissue) gray level in CBCT acquisition with respect to traditional CT. In particular, vessels and bone structure have similar statistical distributions, meaning that a simple threshold cannot properly discriminate between the two tissues.

Based on this observation, the proposed method performs, as first step, a bone segmentation, using not-contrasted images. The resulting segmentation is used as a mask to improve vessel extraction in post-contrast images. Both steps are based on a region growing algorithm, with appropriate inclusion criteria.

## 2    Materials and Methods

The present study introduces a CT segmentation algorithm that is planned to be integrated in the radiological platform produced from the company Epica Imaginalis, based on CBCT technology. The main characteristic that distinguish CBCT technology from traditional CTs consists on a conic geometry of X-rays beam. The conic geometry allows to acquire the entire Field of View (FOV) in each exposition. Thus, in contrast with conventional fan-beam tomography, CBCT acquires a complete volume of the region of interest with a single rotation, instead of multiple spiral rotations.

The proposed 3D segmentation algorithm was developed in Matlab using two data sets of 400 DICOM (Digital Imaging and Communications in Medicine) images per each. These two data sets consist of two tomographic acquisitions that were obtained, respectively, before and after the administration of contrast medium.

The algorithm is a two step procedure; in the first step, segmentation of bone tissue is carried out on the pre-contrast DICOM set. The output of this step is used as a binary mask for eliminating voxels corresponding to bone tissue from post contrast image set; the masked images are used for vessels segmentation as contrast medium augments vessels capacity of absorbing X-rays. In this way vascular enhancement achieves a level that is useful for segmentation, without incurring in the overlapping of grey level distributions in bones and vessels. The schematic diagram of the algorithm is shown in Fig. 1.

Both steps of the algorithm adopt a region growing procedure as main segmentation technique. This approach to segmentation iteratively groups pixels into larger regions based on a proximity growth and according to a predefined similarity criteria. The region growing algorithm is completely defined by the proximity criteria, describing spatial relationships to pixels that are considered for grouping, by the similarity criteria for continuing to grow the region, and by the seed points that are used to initialize the region to be segmented. The algorithm starts from selected seed points, distributed over the image, and appends new pixels, in the neighborhood to the already segmented region, if they satisfy predefined criteria. The low computational cost of this technique and the fact that is based on local properties of the region of interest has made it applicable both for bones and vessels segmentation.

### Bones Segmentation

Bone is a dynamic connective tissue, having a complex structure which is directly related to its mechanical and biological functions. The constituent tissues of bone are cortical bone and cancellous bone [11].

Cortical bone is solid and compact and comprises the majority of adult skeleton. Instead, cancellous bone is comprised of a three dimensional lattice of intersecting boney plates called trabeculae. The spaces between the trabeculae are filled with bone marrow. As a consequence of their composition, cancellous bone presents greater porosity than cortical tissue and is less dense [2]. The different densities of these two tissues determines a wide range of intensities associated to

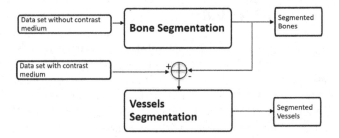

**Fig. 1.** Schematic overview of the proposed algorithm.

bones in CT images. This aspect has deeply influenced the segmentation protocol implemented in this study.

In particular, the proposed procedure is composed by two principal approaches. The first one is based on the region growing technique. This method focuses on the extraction of cortical bone since compact bone has a limited range of variability of the gray levels with a relatively large contrast with surrounding tissues. On the contrary, as spongy tissue has a minor density and often presents discontinuous and large local variations of gray levels, the region growing technique often fails to segment cancellous bone. Therefore, a second procedure step focuses on spongy bone extraction.

**Cortical Bone Segmentation.** Region growing algorithms depend on the identification of the initial seed point. The determination of starting seed points is based on the application of a global threshold $\theta_g$ to the volume obtained from the pre-contrast acquisition, as cortical bone is characterized by the highest intensities in radiological images. Starting from initial seed points, region growing process expands the volume of interest around each voxel $q$, at coordinates $(x, y, z)$ using two criteria of inclusion. The first one implies that the voxel $p$, in the neighborhood $B(x, y, z)$ of $q$, presents a sufficiently elevate grey level $I_p$, as an evidence that it belongs to bone tissue; this condition is expressed as $I_p > \theta_c$, where the threshold value statisfies $\theta_c < \theta_g$. The second criterion determines whether the voxel intensity and that of the seed point lie within a given range: $I_p - I_q < delta$. If the two conditions are satisfied the examined voxel is appended in the list $L$ of seed pixels that are to be analyzed by the region growing. When the list empties the algorithm stops. The pseudo code of the algorithm is shown in Fig. 2.

The method has demonstrated good results for cortical bone segmentation. However, as the region growing approach is based on the difference of intensity between the seed point and the considered voxel, it turned to have limits in the identification of spongy tissue points whose gray levels do not belong to the interval imposed by region growing criteria (as shown in Fig. 3). This aspect implied the development of a specific algorithm (Fig. 4).

```
Initialize: Seed points list L
repeat
     Remove the first voxel q at (x,y,z) from L
     for all voxel p in B(x,y,z)
          if I(q)-I(p)< local threshold and
          I(p)> absolute threshold
               Add p into L
Until L is empty
```

**Fig. 2.** Pseudo code for region growing algorithm.

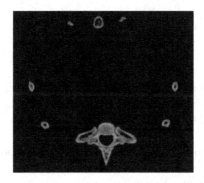

**Fig. 3.** Frame resulting from cortical bone segmentation.

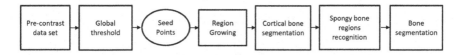

**Fig. 4.** Overview of the presented algorithm for bone segmentation.

**Cancellous Bone Segmentation.** The method introduced for compact bone segmentation presents low performances in the individuation of spongy tissue areas inside of bones; the problem, indeed, is not a segmentation of spongy tissue in strict sense, but rather the classification of the tissue located inside a cavity in the segmented bone tissue. Indeed, bone cavities that are created by region growing approach (as shown in Fig. 3), belong either to cancellous tissue regions or areas corresponding to the vertebral canal, which contains mostly nervous tissue.

Therefore, a dedicated algorithm was implemented in order to classify those cavities as spongy tissue or as nervous tissue. In order to achieve this purpose, a morphological analysis was conduced considering every bone cavity encountered for each subject. For each area, a vector $P_i$ of four parameters was calculated: two morphological parameters (i.e. eccentricity $e_i$ and circularity $c_i$), gray level variance $var_i$ and mean intensity $mu_i$ of the pixels of each region under study; $var_i$ and $mu_i$ provide important information about the examined region in terms of its density, while its similarity with a circular shape is represented by $e_i$ and $c_i$.

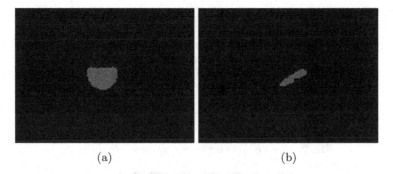

(a)                      (b)

**Fig. 5.** Examples of a bone cavity corresponding to vertebral canal (a) and a spongy tissue region non-segmented by region growing method (b).

The analysis indicates that regions belonging to the vertebral canal are characterized by a greater area than cancellous tissue cavities and present an almost circular shape (Fig. 5); this can be observed in the distribution of eccentricity and circularity in each class (Fig. 6a). Limited variation of variance and mean intensity values of vertebral canal areas, in respect with spongy tissue ones, indicate that cancellous bone is much more porous than nervous tissue as its local density is not homogeneous (Fig. 6b).

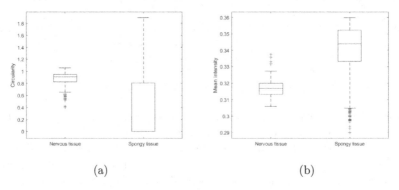

(a)                      (b)

**Fig. 6.** As example are reported the boxplots corresponding to Circularity (a) and Mean Intensity distributions for both nervous tissue and spongy tissue.

The analysis conduced on bone cavities permitted to define the conditions that ensure the discrimination of vertebral canal cavities from spongy tissue holes. Starting from these observations, the second step of the bone segmentation algorithm detects all cavities in each TC slice, and classifies each of them accordingly with a decision tree algorithm, reported in Fig. 7. In particular, $C$ is the vector containing every identified hole, suffixes $Inf$ and $Sup$ represent, respectively, the inferior limit and the superior limit of each parameter, $S$

```
Initialize: vector C containing bone cavities
repeat for i = 1 to length(C)
    Consider parameters e, c, var, mu of i-th bone cavity
    if (e(i) < eInf or e(i) > eSup or
        c(i) < cInf or c(i) > cSup or
        mu(i) < muInf or mu(i) > muSup or
        var(i) < varInf )
            Add i-th cavity to S
```

**Fig. 7.** Pseudo code for spongy tissue classification.

corresponds to the set of spongy tissue cavities. In particular, cavities resulting as spongy tissue regions are annexed to bones volume.

Above-mentioned method succeeds in identifying most part of spongy tissue regions. A second approach (Fig. 8) was developed in order to deal with remaining cancellous bone cavities. It consists on verifying whether the hole found in a certain tomographic slice is present in its adjacent slices too. If no cavities are found in at least one among previous slice and successive one, then the analyzed region is considered as a spongy tissue area.

Results of cancellous bone segmentation are shown in Fig. 9b.

```
Initialize: vector C containing bone cavities
repeat repeat for i = 1 to length(C)
    Calculate the slice corresponding to i-th cavity
    if previous slice has no hole or
       successive slice has no hole
          Add i-th cavity to S
All cavities not yet classified are classified as N
```

**Fig. 8.** Pseudo code for the recognition of spongy tissue areas that do not follow classification criteria.

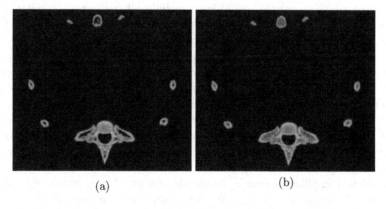

(a)                                      (b)

**Fig. 9.** Results of bone segmentation before (a) and after (b) spongy tissue segmentation.

**Vessels Segmentation**

This section presents the method that performs vessels segmentation. The procedure is based on a data set obtained by a tomographic acquisition with contrast medium. In fact, contrast medium augments vascular enhancement, increasing contrast between vessels and other soft tissues (i.e. muscles). On the other side, the presence of CM increases absorption in the vascular district, whose gray levels become comparable with gray level in bones. Thus, a large improvement in the segmentation procedure can be obtained by using the previously segmented bones as a mask to subtract the bone region from post-contrast data set.

Vessel segmentation also adopts a region growing approach. In this case, the detection of the starting seed point is based on the detection of a circular region, having a suitable diameter and intensity. This circular region is assumed represent one of the large blood vessels. Thus, the center of the located area is then used as the initial seed point. The detection of the circular region is obtained with the Hough transform. Also, in case the selected slice does not provide a suitable seed, the process is repeated in the following slide, until a suitable match is found. Once initialized, the region growing algorithm expands the volume corresponding to vascular district using a predefined criterion of inclusion: the difference of intensity between the candidate pixel and the seed pixel must lie within the specified range. In this case, the algorithm therefore does not impose a predefined intensity range for the blood gray level.

## 3   Results

In order to validate the currently proposed segmentation protocol we compared the obtained results with two semiautomatic methods and a manual segmentation provided by Segment Editor, present in the software *3DSlicer*.

The first semiautomatic method is a threshold segmentation method. The user selected two different threshold ranges, one for bone segmentation and a second one for vessels. The identification of the first (bone) threshold was relatively easy, and the final value is quite good to segment bone structure. On the other side, a single threshold, capable to include all the venous capillaries and excluding all other tissue not of interest (Fig. 10) appeared unfeasible, as it was impossible to obtain an adequate segmentation.

In the second step, we compared our study with the region growing tool, also provided in Segment Editor. The region growing tool allows to segment volume using manual selection of the object of interest and the background. The user has to select object and background in different slices and views, then the user applies the *grow from seeds* tool. This operation is very expensive in term of time and computational cost, and it is also deeply dependent on the user experience. The results indicate that this semiautomatic method performs a good segmentation of the larger vessels, including heart and larger pots, but does not detect capillaries and little vessels. Figure 11 reports the comparison between vessels segmented using *grow from seeds* and our algorithm.

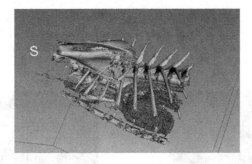

**Fig. 10.** Bones and vessels resulting from the application of *3DSlicer* threshold tool.

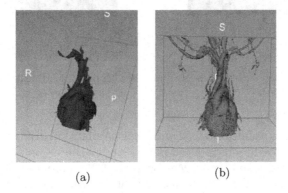

(a)                                 (b)

**Fig. 11.** Vessels volume resulting from region growing module of *3DSlicer* (a) and the proposed segmentation algorithm (b).

Both threshold segmentation and region growing method have low performances in detecting vessels of little dimensions. This problematic is overcome with manual segmentation from the section *hollow*. As shown in Fig. 12, the accuracy of vessels reconstruction, obtained with manual segmentation, is very good and can be used as a reference segmentation for validating our algorithm. However, the manual approach requires a large amount of time and the results are strongly influenced by the clinical knowledge of the user.

In order to validate quantitatively our system, sensitivity and specificity have been calculated. A test will be much more sensitive the lower are the false negative (i.e. pixel mistakenly classified as vessels). This concept stands in contrast with sensitivity, that is, the ability of the test to correctly identify vessels. Pixel identified as vessels in the manually segmented frame are considered as true positive. The test has been performed on a sub set of images mainly containing small vessels, where the segmentation process is more difficult. Each of them has been manually segmented, and results were compared with the output of our segmentation algorithm. Sensitivity has a value of 0.748, and specificity a value of 0.999.

Results suggest that the proposed algorithm is somewhat conservative, i.e. the number of extra pixels classified as vessels is very small, while part of the smaller vessels, and of vessel boundaries, were classified as "not vessels".

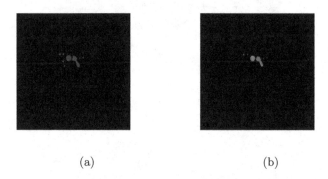

(a)           (b)

**Fig. 12.** Comparison between vessels extracted with manual segmentation (a) and the suggested method (b).

Comparing the results obtained with tools available in 3DSlicer, we can conclude that the procedure introduced in this study achieves an accurate segmentation of both vessels and bones. Moreover, the fact that our protocol is entirely automatic excludes errors due to manual interactions and subjectivity in the segmentation result. In addition, future works will see the implementation of the algorithm in C language. This will improve segmentation protocol speed and facilitate its clinical application.

## 4 Discussion and Conclusions

The presented 3D segmentation algorithm consists on two segmentation stages. The first one is focused on bones extraction. Resulting bones are used as a mask to support and improve the following step of blood vessels segmentation. Bone reconstruction is conduced on CT volume without contrast medium. Instead, vessels extraction is based on CT images acquired after contrast medium injection. The use of contrasted images has the aim of increasing vascular enhancement which is necessary for segmentation success.

Both vessels and bones segmentation rely on region growing technique. These two approaches are associated by the fact that region growing criteria is focused on difference of intensity between seed points and adjacent voxels. However, bone segmentation is strongly influenced by bone composition, which is mostly characterized by cortical tissue and cancellous tissue. Region growing approach succeeds in compact bone segmentation while spongy tissue required the development of a dedicated algorithm that is based on a morphological analysis.

Blood vessels rendering has plenty of possible applications in medicine. However, this study concentrates on achieving 3D blood vessels reconstruction in order to visualize these anatomical models through augmented or virtual reality systems. In order to validate the visualization system the proposed algorithm was used to visualize resulting anatomical structures by means of a virtual reality viewer. Since both vessels and bones volumes were saved as DICOM files, a conversion into a more portable format was necessary in order to facilitate the exportation of 3D surfaces on the virtual reality platform. The selected extension is OBJ format. The conversion in .obj was achieved thanks to a dedicated tool of open source software, *3DSlicer* [8,12].

Figure 13 shows blood vessel and bone anatomical models visualized with a virtual reality platform.

**Fig. 13.** Vessels and bones anatomical models visualized on the available VR system.

# References

1. Bin, S., Masood, S., Jung, Y.: Virtual and augmented reality in medicine. In: Biomedical Information Technology, pp. 673–686. Elsevier (2020). https://doi.org/10.1016/B978-0-12-816034-3.00020-1
2. Black, J.D., Tadros, B.J.: Bone structure: from cortical to calcium. Orthop. Trauma **34**(3), 113–119 (2020). https://doi.org/10.1016/j.mporth.2020.03.002
3. Calllara, A.L., Magliaro, C., Ahluwalia, A., Vanello, N.: Smart region-growing: a novel algorithm for the segmentation of 3D clarified confocal image stacks. Frontiers in Neuroinformatics (2018). https://doi.org/10.3389/fninf.2020.00009
4. Campochiaro, A.: Molecular pathogenesis of retinal and choroidal vascular diseases. Prog. Retinal Eye Res. **49**, 67–81 (2015). https://doi.org/10.1016/j.preteyeres.2015.06.002
5. Carmeliet, P., Jain, R.K.: Angiogenesis in cancer and other diseases. Nature **407**(6801), 249–257 (2000). https://doi.org/10.1038/35025220
6. De Momi, C., et al.: Automatic trajectory planner for stereoelectroencephalography procedures: a retrospective study vascular diseases. Trans. Biomed. Eng. **60**(4), 986–996 (2013). https://doi.org/10.1109/TBME.2012.2231681
7. De Paolis, L.T., De Luca, V.: Augmented visualization with depth perception cues to improve the surgeon's performance in minimally invasive surgery. Med. Biol. Eng. Comput. **57**(5), 995–1013 (2019). https://doi.org/10.1007/s11517-018-1929-6

8.  Fedorov, A., et al.: 3D slicer as an image computing platform for the quantitative imaging network. Mag. Reson. Imag. **30**(9), 1323–1341 (2012). https://doi.org/10.1016/j.mri.2012.05.001

9.  Fu, H., Xu, D., Liu, J.: Retinal vessel segmentation via deep learning network and fully-connected conditional random fields. In: IEEE 13th International Symposium on Biomedical Imaging (ISBI), vol. 16, pp. 695–701 (2016). https://doi.org/10.1109/ISBI.2016.7493362

10. Jones, D., Stangenberg, L., Swerdlow, N.: Image fusion and 3-Dimensional roadmapping in endovadscular surgery. Ann. Vasc. Surg. **52**, 302–311 (2018). https://doi.org/10.1016/j.avsg.2018.03.032

11. Kadir, M.R.A., Syahrom, A., Öchsner, A.: Finite element analysis of idealised unit cell cancellous structure based on morphological indices of cancellous bone. Med. Biol. Eng. Comput. **48**(5), 497–505 (2010). https://doi.org/10.1007/s11517-010-0593-2

12. Kikinis, R., Pieper, S.D., Vosburgh, K.G.: 3D Slicer: a platform for subject-specific image analysis, visualization, and clinical support. In: Jolesz, F.A. (ed.) Intraoperative Imaging and Image-guided Therapy, pp. 277–289. Springer, New York (2014). https://doi.org/10.1007/978-1-4614-7657-3_19

13. Moccia, S., De Momi, E., Sara, E.H., Mattos, L.S.: Blood vessel segmentation algorithms - review of methods, datasets and evaluation metircs. Comput. Methods Programs Biomed. **158**, 71–91 (2018). https://doi.org/10.1016/j.cmpb.2018.02.001

14. Mühlenbruch, G., Das, M., Hohl, C.: Global left ventricular function in cardiac ct. evaluation of an automated 3D region-growing segmentation algorithm. Eur. Radiol. **16**, 117–1123 (2006). https://doi.org/10.1007/s00330-005-0079-z

15. Rogai, F., Manfredi, C., Bocchi, L.: Metaheuristics for specialization of a segmentation algorithm for ultrasound images. IEEE Trans. Evol. Comput. **20**(5), 730–741 (2016). https://doi.org/10.1109/TEVC.2016.2515660

16. Silva, J.N., Southworth, M., Raptis, C., Silva, J.: Emerging applications of virtual reality in cardiovascular medicine. JACC: Basic Trans. Sci. **3**(3), 420–430 (2018). https://doi.org/10.1016/j.jacbts.2017.11.009

17. Tetteh, G., Efremov, V., Forkert, N.D.: DeepVesselNet: vessel segmentation, centerline prediction, and bifurcation detection in 3-D angiographic volumes. CoRR abs/1803.09340 (2018). https://doi.org/10.3389/fnins.2020.592352

# A Model of Pupillometric Signals for Studying Inherited Retinal Diseases in Childhood Population

Daniele Ermini[1] , Rachele Fabbri[1] , Ernesto Iadanza[1] ,
Monica Gherardelli[1] , Paolo Melillo[2] , and Leonardo Bocchi[1(✉)]

[1] Department of Information Engineering, University of Florence, Via S. Marta 3,
50139 Florence, Italy
leonardo.bocchi@unifi.it
[2] Eye Clinic, Multidisciplinary Department of Medical, Surgical and Dental Sciences,
University of Campania Luigi Vanvitelli, 80100 Naples, Italy

**Abstract.** A cutting-edge method to assess the status of photoreceptors is represented by chromatic pupillometry, which consists in stimulating the eyes of the patients with light of different wavelengths and intensities. The signals of the dynamics of the pupil after the optical stimulations can be approximated by a 2nd-order linear model. Fitting parameters are established by a weighted least-squares algorithm, in order to achieve the correct estimate of the shape of the pupillometric data. Results presented in this work indicate that our model is able to capture the dynamics of pupillometric signals.

**Keywords:** Pupillometry · Modeling · Pupil dynamics

## 1 Introduction

With Inherited Retinal Diseases (IRDs) we indicate an heterogeneous set of ocular disorders causing severe visual deficit or blindness [9]. IRDs can be classified as outer retina (i.e. photoreceptors degenerations) and inner retina diseases (i.e. retinal ganglion cell degeneration). In both cases, the disease is characterised by a very high heterogeneity in the involved genes. Such a high heterogeneity represents an obstacle in making a clinical evaluation, in order to make a diagnosis and establish an appropriate treatment. Moreover, the diagnosis method are usually based on invasive techniques like electroretinography (ERG), which can be particularly difficult to be applied on a children.

In this scenario, chromatic pupillometry represents a novel method to assess the status of the photoreceptors, rods and cones (i.e. retinal cells) [6,7], and it is particularly recommended for children because of its low invasiveness. Chromatic pupillometry consists in the stimulation of the patient's eyes with light of different wavelength and intensities. Indeed, different conditions of the stimulation cause a selective response of rods and cones and allow to evaluate their functionality.

© Springer Nature Switzerland AG 2021
J. Hasic Telalovic and M. Kantardzic (Eds.): MeFDATA 2020, CCIS 1343, pp. 69–78, 2021.
https://doi.org/10.1007/978-3-030-72805-2_5

The data and the results presented in this paper have been collected within a project funded by the Italian Ministry of Education and Research. A pilot study has been designed, involving both clinicians and engineers working in different Italian Universities. The multidisciplinary team has worked with the goal of creating protocols and systems able to perform an early diagnosis and assessment of IRDs, and in particular of Retinitis Pigmentosa (RP), in a pediatric population through chromatic pupillometry.

The activity has already brought to some important results. First, a protocol for the stimulation has been designed. It establishes the characteristics of the light in terms of color and intensity and also of the background [8]. Then, an Information Technology (IT) web platform has been developed in order to gather and share the data collected by the clinical centers. This platform has been named ORÁO which stands for Ophthalmologic medical Record for chromAtic pupillOmetry. The acronym is also a verb from the ancient greek, meaning *I see*. The platform has been designed with a RESTful architecture and it is web based, thanks to the cloud insfrastructure provided by the Distributed Internet and Technology (DISIT) Lab of the Department of Information Engineering of the University of Florence [1]. Further details of the design of ORÁO can be found in [3] and [4].

The platform is powered with a Clinical Decision Support System (CDSS) based on Machine Learning (ML) techniques. The CDSS exploits Support Vector Machines (SVMs) trained with the pupillometry parameters as input to predict whether the patient is affected by RP or not. The CDSS is fully described in [5].

One of the last project steps has been the preliminary study of a model-based approximation of the dynamics of the pupil diameter after the light stimulations provided according to the protocol of the research. The preliminary results have been presented in [2]. However, the procedure has been further optimized and the model has been improved. In this paper, the latest results are shown.

## 2  Materials and Methods

### 2.1  Dataset

The dataset used in this work is composed of 182 subjects; 31 are affected by RP, labeled as *Patients*, 64 healthy subjects, labeled as *Controls* and 87 subjects without a label. The data comes from the database of the above mentioned Electronic Medical Record, ORÁO, which collects the ophthalmological and pupillometric data.

The pupillometric response has been analyzed for every subject and the signals of the pupil diameter over time have been stored in a .DAT file. The protocol for the acquisition procedure has been described in [3–5]. In order to reduce the noise, a FIR filter was used for each pupil response to light stimulation, as described in [5]. For each subject, 40 stimulations are available (half respectively for left and right eye). Therefore, a total of 7,280 pupillometric signals has been used for the modelling procedures described in this paper.

## 2.2    Model for the Signal

The state of art does not provide a mathematical model that describes the exact dynamics of the pupil diameter. However, the experimental shape of the curve suggests that a linear approximation of the system could deliver reasonable results. Therefore, we adopted the model used in [2] and [10] that is able to replicate the observed shape of the pupillary reaction to a light stimulus; the simplest way to realize such shapes is a second order linear time invariant model. The impulse response, produced by an impulse input at time $t = t_0$, can be expressed as:

$$f(t) = \begin{cases} k - A \left( e^{-\frac{t-t_0}{\tau_d}} - e^{-\frac{t-t_0}{\tau_r}} \right) & \text{if } t > t_0 \\ k & \text{otherwise.} \end{cases} \quad (1)$$

The parameters of the model are:

- $k$ is the pupil diameter at rest;
- $t_0$ represents the starting time of the pupil response that could be delayed;
- $A$ is related to the maximum difference between the diameter values during the stimulation;
- $\tau_d$ is the constant time of the constriction phase of the pupil;
- $\tau_r$ represents the constant time of the phase of returning to the diameter rest.

## 2.3    Parameter Estimation, Fitting and Optimization

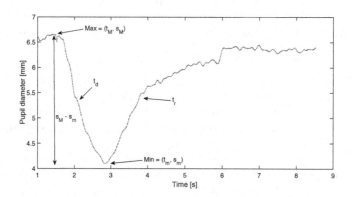

**Fig. 1.** Parameters estimation procedure.

A procedure to estimate properly the vector $\hat{p} = (\hat{k}, \hat{t}_0, \hat{A}, \hat{\tau}_d, \hat{\tau}_r)$ of the model parameters is described in [2]. It begins by removing all possible outliers and locating the first two fiduciary points: the maximum and the minimum of the pupil diameter, labeled respectively as $(t_M, s_M)$ and $(t_m, s_m)$. The value $s_M$ is an estimate for the parameter $\hat{k} = s_M$, which is the rest length of the pupil. The amplitude A is estimated by the difference $\hat{A} = s_M - s_m$. $t_d$ and $t_r$ represents the halving time where the signal reaches the value $s(t) = s_m + (s_M - s_m)/2$. The halving times are used to estimate the parameters $\hat{\tau}_d$ and $\hat{\tau}_r$: $\hat{\tau}_d = ln(2)(t_m - t_d)$ and $\hat{\tau}_r = ln(2)(t_r - t_m)$. The starting time $\hat{t}_0$ is then estimated with the relation $\hat{t}_0 = s_M - 0.2(s_M - s_m)$ (Fig. 2).

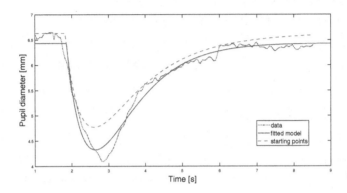

**Fig. 2.** Experimental signal, model with parameters estimation and after Weight Least Squares algorithm.

These estimations reasonably approximate the experimental data. However with a fitting process it is possible to achieve a better output, reducing the error between the data and the fitting. Therefore $\hat{p}$ is used as an initial estimate for the iterative Weighted Least Squares (WLS) algorithm, in order to obtain the optimal values of the parameters $\tilde{p} = (\tilde{k}, \tilde{t}_0, \tilde{A}, \tilde{\tau}_d, \tilde{\tau}_r)$. The variability range applied on the parameters are defined in [2].

The duration of the pupil response is significantly shorter than the time window of the measurements. Thus we decided to use only the first 6.5 s of the signal and to consider the 0.5 s prior to stimulation in order to highlight the rest phase at start time. This procedure reduces the number of artifacts, which can disturb the WLS algorithm, and leads to better results in terms of Root Mean Square Error (RMSE) and R-Squared Coefficient.

The weight distribution we adopted has a trapezoidal distribution as shown in Fig. 1. The samples of the prior stimulation (16 samples) have the minimum values; the following 100 samples have the maximum value because the first seconds of the response contain the most informative part of the signal; to the remaining samples correspond a linear decrease of weights.

The signal in each time window of interest has been independently fitted with the model and the R-Squared Coefficient has been used as an evaluation of the fit efficacy (Fig. 3).

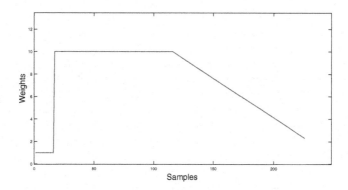

**Fig. 3.** Distribution of the weights for each sample.

## 3    Results

The preliminary analysis of the parameter plots as a function of R-Squared coefficient lets us understand when the fitting procedure is not able to cover the dynamic of the measured signal. The parameters distribution as a function of the R-Squared coefficient is shown in Fig. 4. The precision of the fit increases with higher R Square values.

A visual inspection of the output of the WLS algorithm reveals that fittings with a low R-Squared coefficient are not usable. Their presence can be explained with input data noise. Indeed, the acquisition process can be jeopardized by eye-blinkings, head-movements or similar. These are due to the uncomfortable status of the subject after several stimulations. A selection step was introduced to remove the bad fittings from our analysis. We discarded all the fitting results with a R-Squared coefficient lower than 0.7 or with a value of the K parameter grater than 9 mm as it has no physiological significance. This process extracts 5,556 valid models from the full dataset with a 76.3% of good fits.

Subsequently, we focused on the statistical difference between the parameters distribution for the *Controls* and *Patients* data. The aim of this phase is to evaluate if the output of the model is significantly different between the labels, i.i. between *Controls* and *Patients*. The distribution of the parameters K, A and $\tau_d$ is shown in Fig. 5 in form of boxplot. Note that the dimension of the labels is different and for the $\tau_d$ parameter the plot contains some outlier not visible with the selected scale. A visual analysis suggests that can be a statistical difference between the labels in some parameters. A Wilcoxon Rank Sum test at 5% significance has been performed in order to analyze the median distribution of the parameters K, A and $\tau_d$, assuming the hypothesis of independence among the labels. Null hypothesis consists in assuming that data are samples from continuous distributions with equal medians, against the alternative hypothesis that they are not. Results are reported in Table 1.

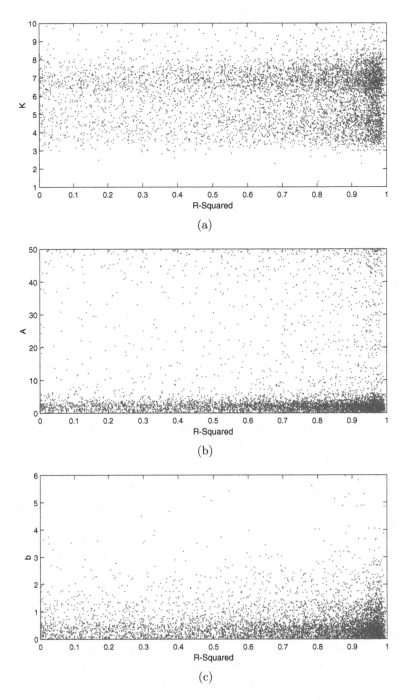

**Fig. 4.** Plots of the parameters K (a), A (b), $b = 1/\tau_d$ (c), $c = 1/\tau_r$ (d), $t_0$ (e) as a function of the R-Squared coefficient.

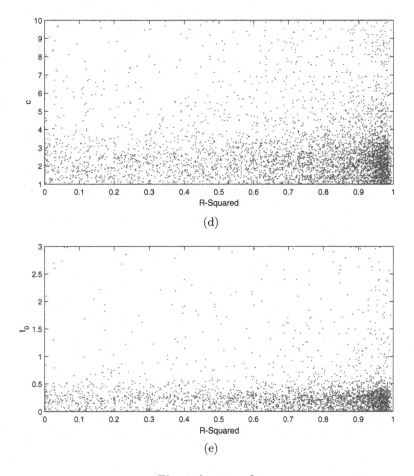

**Fig. 4.** (*continued*)

**Table 1.** Results of the Wilcoxon Rank Sum test on the parameters K, A and $\tau_d$.

|          | k                      | A                       | $\tau_d$             |
|----------|------------------------|-------------------------|----------------------|
| p-value  | $6.65 \times 10^{-16}$ | $65.23 \times 10^{-9}$  | $0.37 \times 10^{-2}$ |

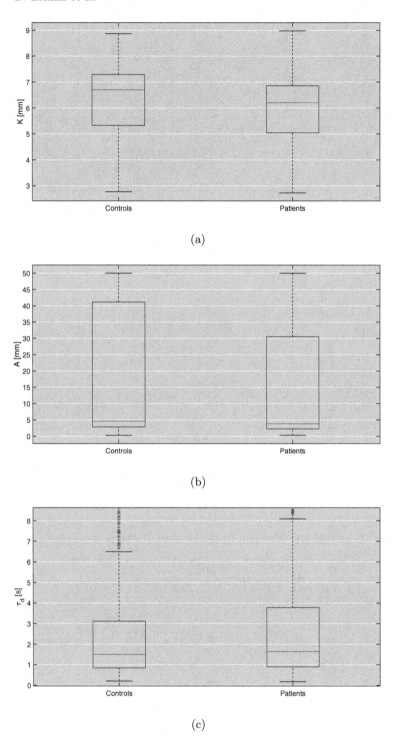

**Fig. 5.** Boxplots of the parameters K (a), A (b), $\tau_d$ (c)

# 4   Discussion and Conclusions

Data shown in previous sections shows that the proposed model is able to approximate the pupillometric signal. However we have to consider that there is a significant number of samples failing the fit procedure. An improvement of the quality of the acquisition process could be a possible way to reduce artifacts and therefore avoid operating with unusable data. The statistical analysis of the parameters based on the Wilcoxon Rank Sum test reveals that a different distribution of the labels *Controls* and *Patients* exists. Values of $p < 0.05$ allow us to reject the null hypothesis of equal medians of the data.

As future development, this analysis would be extended, focusing on the different wavelengths and intensities of the light stimulus with the aim of defining the optimal stimulus, in terms of statistical difference among *Controls* and *Patients*. Another possible development for this work could be also creating a classifier which exploits the parameters of the models as classification features. It would be interesting to evaluate the differences of performance, compared with the already existing classifier system (included in the CDSS mentioned in Sect. 1) which works directly with the output data of the pupillometer. Another relevant topic to work on would be the study of a classifier that uses an ensemble approach for the pupillometric analysis.

# References

1. Bellini, P., Bruno, I., Cenni, D., Nesi, P.: Managing cloud via smart cloud engine and knowledge base. Future Gener. Comput. Syst. **78**, 142–154 (2018)
2. Fabbri, R., Iadanza, E., Gherardelli, M., Melillo, P., Simonelli, F., Bocchi, L.: Modeling of pupillometric signals for studying children's rare diseases. In: Asian Pacific Conference on Medical and Biological Engineering. Springer (2020, in press)
3. Iadanza, E., et al.: ORÁO: RESTful cloud-based ophthalmologic medical record for chromatic pupillometry. In: Badnjevic, A., Škrbić, R., Gurbeta Pokvić, L. (eds.) CMBEBIH 2019. IP, vol. 73, pp. 713–720. Springer, Cham (2020). https://doi.org/10.1007/978-3-030-17971-7_106
4. Iadanza, E., Fabbri, R., Luschi, A., Melillo, P., Simonelli, F.: A collaborative restful cloud-based tool for management of chromatic pupillometry in a clinical trial. Health Technol. **10**(1), 25–38 (2020). https://doi.org/10.1007/s12553-019-00362-z
5. Iadanza, E., et al.: Automatic detection of genetic diseases in pediatric age using pupillometry. IEEE Access **8**, 34949–34961 (2020)
6. Kardon, R., Anderson, S.C., Damarjian, T.G., Grace, E.M., Stone, E., Kawasaki, A.: Chromatic pupillometry in patients with retinitis pigmentosa. Ophthalmology **118**(2), 376–381 (2011)
7. Kawasaki, A., Collomb, S., Léon, L., Münch, M.: Pupil responses derived from outer and inner retinal photoreception are normal in patients with hereditary optic neuropathy. Exp. Eye Res. **120**, 161–166 (2014)
8. Melillo, P., et al.: Chromatic pupillometry for screening and monitoring of retinitis pigmentosa. Invest. Ophthalmol. Vis. Sci. **60**(9), 4513 (2019)

9. Smith, J., Ward, D., Michaelides, M., Moore, A., Simpson, S.: New and emerging technologies for the treatment of inherited retinal diseases: a horizon scanning review. Eye **29**(9), 1131–1140 (2015)
10. Tarchi, P., et al.: Modeling of voltage imaging for the study of action potential propagation. In: Badnjevic, A., Škrbić, R., Gurbeta Pokvić, L. (eds.) CMBEBIH 2019. IP, vol. 73, pp. 309–314. Springer, Cham (2020). https://doi.org/10.1007/978-3-030-17971-7_47

# The Use of Data Science for Decision Making in Medicine: The Microbial Community of the Gut and Autism Spectrum Disorders

Jasminka Hasic Telalovic$^{(\boxtimes)}$ ⓘ, Lejla Pasic ⓘ, and Dzana Basic Cicak ⓘ

University Sarajevo School of Science and Technology, 71000 Sarajevo, Bosnia and Herzegovina
jasminka.hasic@ssst.edu.ba

**Abstract.** The study of microbiome composition is showing positive indication for the utilization in the diagnosis and treatment of many conditions and diseases. One such condition is an autism spectrum disorder (ASD). In this paper, we analyzed a dataset of 58 children's gut microbiome 16 s rRNA sequenced samples from Ecuador. The 31 samples in the dataset were from the neurotypical children (Control group), while 27 were from the children with ADS symptoms. We analyzed the dataset and reported the most statistical significance between the two studied groups. Furthermore, we applied the Random Forest (RF) machine learning algorithm to develop a classifier that distinguishes neurotypical samples from the ASD samples. The features of the classifier were relative abundances of genus-level bacteria in each sample. The best performance of the classifier (with 5-fold cross-validation) was exhibited when the top five features were used (as identified by RF feature importance metric). The overall accuracy was (83.3%). The ASD samples were identified with 75% accuracy while neurotypical samples were identified with 87.5% accuracy.

**Keywords:** Machine learning · Random forest · Microbiome · ASD · 16 s rRNA sequencing

## 1 Introduction

Over the past two decades, the development of novel sequencing technologies resulted in the generation of a large number of datasets that correspond to human states of both health and disease. Traditionally, such datasets have been analyzed using statistics to draw population inferences. More recently, they have been used as learning data to find general predictive patterns through machine learning in order to aid medical decision making.

Medical decision making through machine learning showed promise in conditions of complex etiology, which result from a combination of environmental and genetic factors. These include obesity, diabetes mellitus, irritable bowel disease, Crohn's disease, colorectal cancer, and multiple sclerosis [8, 30].

Among these medical conditions is an autism spectrum disorder (ASD). This neurodevelopmental and neuropsychiatric disorder is characterized by onset at a young age,

© Springer Nature Switzerland AG 2021
J. Hasic Telalovic and M. Kantardzic (Eds.): MeFDATA 2020, CCIS 1343, pp. 79–91, 2021.
https://doi.org/10.1007/978-3-030-72805-2_6

deficits in social functioning, cognitive inflexibility, and repetitive behaviors [2]. Over previous decades, the increase in ASD awareness, recognition, diagnosis, and treatment has received substantial attention and its global prevalence has been estimated at over 1% [25].

Pathophysiology of ASD results from interactions between genetic and environmental factors. So far, several hundred genes have been identified as involved in ASD development, a number of which regulate synaptic function [7]. Mutations in any of these genes, their duplications, presence of single-nucleotide polymorphisms, and/or epigenetic changes that affect gene activity and expression, increase the risk of ASD development [17]. Estimates of heritability of the familiar risk of ASD range from 50–83% [23].

Environmental factors involved in ASD development are equally heterogeneous, contribute to up to 40–50% of variance, and include parental age, delivery options, conditions at birth, birth spacing, birth order/parity, gestational age, infections, maternal use of medications, exposure to heavy metals, air pollution and environmental toxins [16].

ASD is known to be associated with a number of medical comorbidities, which include immune dysregulation and up to four times higher prevalence of general gastrointestinal problems [13]. The physiology of the intestine and its associated immune function are regulated by gut microorganisms, collectively known as gut microbiome [27]. Another important role of the gut microbiome concerns the regulation of host behavior, which is affected in ASD. This is achieved through bacterial production of neuro-active compounds that influence the exchange of biochemical signals between the gut and the central nervous system.

The gut microbiome is populated by a large number of different microorganisms that differ in function and abundance. They are typically analyzed by extracting total microbial DNA from feces, followed by amplification of 16S rRNA genes and their mass sequencing using novel sequencing technologies. Obtained datasets are then analyzed using a bioinformatic pipeline to determine the taxonomic composition and the structure of the microbiome. This analysis is based on binning the sequences based on their similarity to operational taxonomic units (OTUs). They correspond to individual bacterial genera and serve as features in further analysis that is applied in this study. Given the ease of this process and the continuing lowering of its costs, a vast amount of such data has become available in previous years.

An important aspect of ASD is its diagnostics, which remains complicated. To meet current diagnostic criteria, the children must present deficits in social-emotional reciprocity, nonverbal communicative behaviors used for social interaction, and deficits in developing, maintaining, and understand relationships. These are assessed through the application of a number of screening questionnaires and are prone to bias, particularly regarding sex/gender differences in the clinical presentation [29]. In addition, these procedures include long-term observation which often results in a substantial time gap between the observed onset of the disorder and diagnosis.

Early diagnosis of ASD is connected to a better outcome as it allows early intervention. In a recent literature review which included fourteen studies, the early intervention approach was viewed as mostly positive with respect to ASD outcome while its value in providing support services to caregivers was universally praised [26]. The motivation for our study is to develop a tool based on the analysis of children's gut microbiome, which can aid the early diagnosis of ASD.

In this study, we have analyzed gut microbiome datasets originating from Ecuadorian children with ASD and neurotypical children. We have used a bioinformatic pipeline to detect different OTUs and statistical analysis to detect and illustrate differences between the two datasets. The obtained list of OTUs was then used as feature input to develop a machine learning classifier that, with high confidence, can distinguish an unknown sample as originating either from a neurotypical child or a child with ASD. To our knowledge, this is the first time that machine learning has been employed as a decision-making tool in the field of gut microbiome analysis of ASD.

## 2 Background

### 2.1 Related Work

Over the previous decade, the approach of gut microbiome sequencing and statistical analysis of obtained data, had demonstrated that individuals with ASD have altered gut microbiota compared to their neurotypical counterparts. Some of the differences detected decrease in Bacteroidetes/Firmicutes ratio, a decrease in *Prevotella*; and an increase in *Sutterella*, *Lactobacillus* and *Desulfovibrio*. However, no uniform findings had been reported as studies were conducted in different parts of the world and microbial composition of the human gut is influenced by many factors which include cultural and nutritional habits [26].

Consequently, probiotic treatments and fecal transplantation emerged as promising novel therapeutic treatments to improve gastrointestinal and behavioral problems in ASD. Recently, it had been reported that fecal transplantation had led to an improvement of the above symptoms in children, which persisted for two years after treatment completion [11].

Machine learning algorithms have been used in gut microbiome research in attempts to improve screening and diagnostic processes. Several studies on the subject have been published and their overview is presented below. Most of these studies used OTUs as features and one was complemented with the usage of k-mers (i.e. nucleotide subsequences of length k). A scoping review of the current research that applies machine learning techniques to microbiome data can be found in Iadanza et al. [10].

Pasolli et al. [18], have analyzed six datasets that corresponded to five conditions: liver cirrhosis, colorectal cancer, IBD, obesity, and type II diabetes mellitus. Four different machine learning algorithms were evaluated: random forest, support vector machines (SVM), elastic net, and LASSO, and were found to have good disease-prediction capabilities.

Reiman et al. [21] used a deep learning algorithm convolutional neural network (CNN) to analyze datasets originating from three body sites which included the gut. The use of CNN showed promise in the classification of body site origin of different metagenomes.

A study by Ai et al. [1] examined gut microbiota-based prediction in colorectal cancer using the base net, random forest, and logistic; and found that these predictions were superior to the standard fecal occult blood diagnostic test.

Asgari et al. [4] explored a combination of deep learning methods (Multi-Layer-Perceptrons neural network) in addition to SVM and Random Forest to identify microbiomes in terms of body site origin and the presence of Chron's disease. They found that the usage of k-mers outperformed OTU approach.

Hasić Telalović and Musić [8] demonstrated that gut microbiomes of typical and multiple sclerosis patients can be classified using a Random Forest classifier. Furthermore, the classifier was validated on an independent dataset and exhibited comparable performance.

All the studies that have built classifiers distinguishing typical from disease samples have reported the Random Forest classifier to have the best performance. Thus, this algorithm was chosen to analyze our dataset.

## 3   Methodology

### 3.1   Dataset

We obtained the dataset from the SRA archive of GenBank (https://trace.ncbi.nlm.nih.gov/Traces/sra/?study=ERP109374). It was generated using Illumina MiSeq sequencing of the conserved V4 region of the 16S rRNA gene. The dataset was comprised of 58 gut microbiome samples that corresponded to 31 neurotypical children (Control group) and 27 children with ASD. All children were from Ecuador and their age varied from zero to thirteen years. Within the neurotypical group, the female to male ratio was 2:29 and within the ASD group it was 1:24. The remaining two samples did not have the sex attribute and their age was specified as zero years. As such, they were explored in the data cleaning phase.

### 3.2   Obtaining Bacterial Abundances

As a first step, the two samples that were discovered to have incomplete and questionable metadata were removed from the analysis. This left 56 samples corresponding to 31 neurotypical and 25 children with ASD.

The sequenced data contained in each sample was analyzed using a bioinformatics pipeline. This analysis determined the bacterial abundances in each sample. The bioinformatics pipeline was developed using the QIIME 2 software tools (version 2020.2) [5] and applied to the paired-end reads in the dataset samples. In this phase all samples exhibited a satisfactory quality for the further analysis.

Sample processing was performed in the following stages:

1. Data import,
2. Dada2 sequence denoising,

– This method denoises single-end sequences, dereplicates them, and filters chimeras.

3. Classification of the reads using classify-sklearn method.

In this stage, a pre-fitted sklearn-based taxonomy classifier is used to classify reads by taxon.

It is important to note that the taxonomy assignment was augmented using a hand-curated SILVA_132 classifier [24]. Namely, unidentified sequences at the species level from the original classifier were blasted against the NCBI database to check for possible improvement following the 99% sequence matching. In the end, 110 k sequence names were altered.

The final product of the bioinformatics analysis resulted in the identification of absolute counts of different bacteria found in each sample. For the purpose of the further analysis, these counts were translated into relative abundances - the amount of each bacteria is expressed as a percentage of the total amount of the bacteria (Fig. 1).

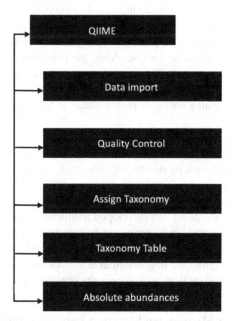

**Fig. 1.** Stages in the bioinformatics analysis performed in this study.

The relative abundance of bacteria was defined at taxonomic levels of phylum, order, family, and genus. These were candidate features for the ML classifier. Based on the results in [8], we chose to use taxa abundance at the genus level for the classifier's features.

For the purpose of statistical description of the dataset, average values of relative abundances in the dataset were calculated and represented as percentage of total. Non-parametric t-test with unequal variance was performed to determine statistically significant differences between the datasets. To calculate alpha-diversity indices Observed species, Shannon, and Simpson, packages Microbiome R [15] and phyloseq [14] were used within R [20].

### 3.3 Random Forest Classifier

We used machine learning for the development of the classifier that classifies an unknown microbiome sample to the two groups (either ASD or neurotypical).

The ML analysis was performed using the Random Forest (RF) classification algorithm. RF is a supervised learning approach and it randomly generates multiple decision trees based on the sample data [12]. The randomness of the bootstrapping of the samples used when building trees, and the sampling of the features to consider when looking for the best split at each node, was set to a fixed number so as to guarantee that the same sequence of random numbers were generated each time the code is run. Each of these trees generates a prediction that counts as a vote. All the votes are evaluated to designate the optimal solution. In addition, an acceptance metric is calculated that expresses feature importance. The feature importance can be used for the model revision: too many features in the model can result in model inaccuracy, some features might not be useful for classification and others might be redundant. For successful classification, number of samples used should be larger than the number of features.

As this a randomized algorithm, different runs of the algorithm may give a different final class assignment to the same sample. Also, the performance may depend on the choice of training and testing data. To avoid these performance biases, we performed 5-fold cross-validation and generated 100 random decision trees. The increase in the number of decision trees may increase performance at the cost of time and memory used.

The accuracy of the model was calculated by the comparison between the model's predicted target data and the actual target data (meaning, if an ASD and neurotypical samples were identified as such by the model). We firstly trained the RF classifier using all the identified genera in our samples as features. We used the obtained model to identify the values of feature importance for each genus taxa. These values were then used to restrict the feature dataset. We developed additional three classification models that use the top three, five, and ten features and assessed their accuracy.

It is an established practice in microbiome studies to report statistical analysis of the studied dataset. In this section, we first summarize the statistical data obtained from the dataset analysis and then we follow with the results obtained in the ML study.

### 3.4 Differences in Gut Microbiome Structure Between Children with ASD and Neurotypical Children as Inferred by Statistical Analysis of Datasets

To understand the differences between the two study groups, we compared the average relative abundances of the OTUs within ASD and Control datasets and determined the statistical significance of observed differences. In Fig. 2 we report the average relative abundances of ten most abundant OTUs in individual datasets, with remaining OTUs grouped as 'Others'. Figure 2a shows that ASD dataset has comparatively higher abundance of OTUs *Bacteroides* (15.1% vs. 11.8%), *Faecalibacterium* (8.6% vs. 6.9%), and *Ruminococcus* (6.4% vs 2.4%) than Control dataset. However, only the difference in abundance of *Ruminococcus* was statistically significant ($p = 0.05$). In addition, *Bilophila*, an OTU of low abundance, was comparably overrepresented in ASD dataset (0.11% vs. 0.05%; $p = 0.03$). Finally, we calculated the ratio between phyla Firmicutes and Bacteroidetes which was 2.2 in ASD and 2.6 in Controls. Thus, while statistical methods show that children with ASD have different structure and composition of gut microbiomes than neurotypical children, they have only limited power to discriminate between these two datasets.

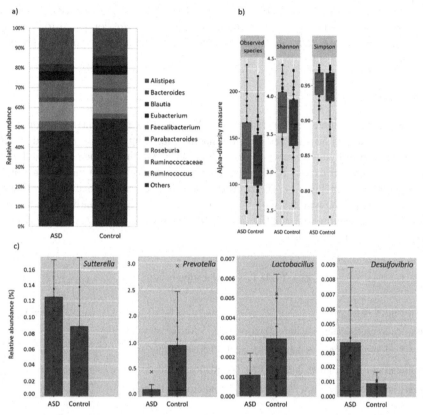

**Fig. 2.** (a) Relative abundance of ten genera in studied datasets; (b) values of alpha-diversity measures Observed species, Shannon index, and Simpson index in individual samples of studied datasets. Boxplots show median with interquartile range, outliers are represented as dots; (c) boxplots of relative abundance of gut microbiome genera previously described as involved in development of ASD. Boxplots show median with interquartile range. ASD-autism spectrum disorder; Control – neurotypical.

Alpha-diversity is a standard diversity measure which describes a diversity within a particular sample. In order to estimate whether the two datasets are populated by different numbers of bacterial genus-level OTUs, we have calculated the values of three alpha-diversity indices (Fig. 2b). Observed species index counts the number of OTUs in individual samples and was higher in children with ASD than in Controls. Shannon index is a more complex alpha-diversity measure that considers OTUs abundance and evenness of their distribution. Its value was also higher in children in ASD compared to Control dataset. Finally, Simpson's index measures the probability that two individuals randomly selected from a sample will belong to the same OTUs. From this analysis, we could conclude that children with ASD have gut microbiomes that are populated by a larger number of bacterial species than their neurotypical counterparts.

We then determined whether the relative abundances of genera previously identified as important in development of ASD follow the pattern seen in previous studies (Fig. 2c). In line with previous studies, ASD dataset showed an increase in abundance of *Sutterella* and *Desulfovibrio* and a decrease in abundance of *Prevotella*. compared to Controls. However, in this dataset, *Lactobacillus* was underrepresented, which contrasts previous findings.

### 3.5  Results

The original dataset contained 295 features (OTUs/genus-level bacteria) and the analyzed dataset set contained 56 samples. For the machine learning analysis, we used the RF implementation in Python's scikit-learn package [19]. We first trained the classifier on this initial set and, as expected, discovered that this would leave to the poor accuracy (54.7% as reported in Table 1). The feature importance scores obtained in this run were utilized to identify the features with the highest importance. The relative abundance of top six features is summarized in Fig. 3. We chose to visualize those six features as they had significantly higher scores.

**Table 1.** Classification results

| Classification approach | Accuracy |
| --- | --- |
| Classification full-features features dataset (RF): 5-fold cross validation | 54.7% |
| Classification-top 3 features dataset (RF): 5-fold cross validation | 72.5% |
| Classification-top 5 features dataset (RF): 5-fold cross validation | 83.3% |
| Classification-top 10 features dataset (RF): 5-fold cross validation | 77.2% |

The top features were used to develop additional 3 ML classifiers. They used the top 3, 5, and 10 features. This choice for a number of features was estimated using the feature importance scores and also results from previous similar studies. The accuracy metric for those classifiers is reported in Table 1. Using fewer features significantly improved accuracy. All three studied models obtained higher accuracy than the model that used all the features. The highest accuracy was obtained by the model that used the top five features (83.3%). The accuracy of the studied groups' identification varied for this model. The ASD samples were identified with 75% accuracy while neurotypical samples were identified with 87.5% accuracy. The significance of these results in discussed in the following section.

As previously noted, we performed 5-fold cross-validation which means that our reported performance is an average performance of multiple classifier runs in which different data in the dataset is designated as testing and training data. To understand better what this accuracy means we report a heatmap for a single such run in Fig. 4.

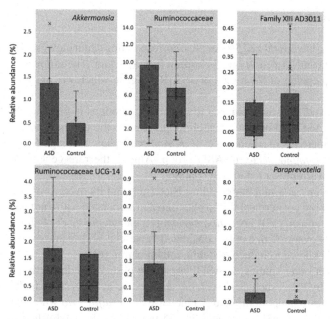

**Fig. 3.** Boxplots of relative abundance the top six most important features showing median with interquartile range. ASD-autism spectrum disorder; Control – neurotypical.

The values on the diagonal correspond to the correctly identified neurotypical samples (7) and ASD samples (3). Those are the true negatives and true positives. The other two values correspond to the incorrectly identified neurotypical (1) and ASD (1) samples. Those are the false positives and false negatives.

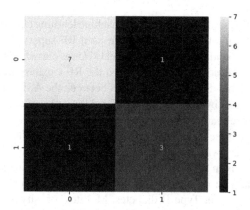

**Fig. 4.** Confusion matrix heatmap showing the number of samples that were correctly and incorrectly classified for each group.

## 4 Discussion

In this paper we have analyzed 56 samples, corresponding to gut metagenomes of 25 children with ASD and 31 neurotypical children from Ecuador. The samples were obtained by MiSeq sequencing of V4 region of 16S rRNA gene which was amplified from total microbial DNA.

We employed statistical analysis to visualize alterations in gut microbiota between children with ASD and neurotypical children. Among the ten most abundant taxa, we observed an increase of abundance of *Faecalibacterium* and significant increase of abundance of *Ruminococcus* in ASD group. Among the rare genera, *Bilophila* was significantly overrepresented in ASD dataset.

*Ruminococcus* is a mucin-degrading bacterium and its overgrowth was associated with an increase in gut permeability, a phenomenon associated with a number of diseases, which include autism [28]. *Faecalibacterium* is a known producer of anti-inflammation butyrate, so its overgrowth can be considered beneficial [22], while overrepresentation of *Bilophila* is known to induce systemic inflammation [6].

In line with other studies, in ASD dataset we observed a decrease in bacteria that degrade carbohydrates such as *Prevotella*; an increase in *Sutterella* which regulates the integrity of intestinal epithelium, and a increase in *Desulfovibrio* which was associated with severity of ASD [3]. *Lactobacillus* did not follow the established pattern, as its abundance was lower in ASD than in Control dataset.

Finally, we observed a decrease in Bacteroidetes/Firmicutes ratio in ASD dataset compared to Control. This finding is commonly observed in obesity, diabetes, ASD and a number of other conditions [27].

In terms of differences in alpha-diversity, ASD dataset had higher diversity of microbial genera than Control dataset. Although this phenomenon was previously observed, up to date no consistent patterns were established when comparing alpha-diversity in neurotypical and children with ASD. Thus, the putative role of these changes remains to be elucidated [9].

In second part of this study, we performed machine learning classification of the ASD and neurotypical gut microbiome samples. We used RF supervised machine learning algorithm and relative abundances of genus-level OTUs as classifier's features. We firstly ran RF on with all the discover OTUs and used the RF's feature importance metric to identify those features that discriminate the best between the ASD and Control samples. We continued our analysis by developing additional three classifiers (using the top three, five, and ten importance features). As expected, the accuracy of classifiers with limited features was higher than the accuracy of classifier that used all the available features.

The highest accuracy was recorded for the classifier that used the top five features (83.3.%). This finding is in line with the accuracies of RF applications to the analysis of gut microbiome data for other medical conditions (liver cirrhosis, colorectal cancer, inflammatory bowel disease, Type II diabetes, Mellitus, obesity, Crohn's disease, and Multiple sclerosis [30] that reported values between 80% and 86%. Our model still has a significant probability of false positive and false negative identification of samples. The ADS diagnosis is a complex process and cannot be done by a sole technique. The timely diagnosis, which allows early intervention, is difficult to achieve using clinical judgment and clinical testing is in its infancy. The accuracy of our methods signifies that it cannot be applied solely for the medical decision making of ASD diagnosis. But, our results are important as they can be used, in the combination with the other available techniques, to help aid early ASD diagnosis.

The microbiome is known to be affected by many different factors, such as diet, medications, lifestyle, stress, different medical conditions, and environmental factors. For validation purposes, it would be desirable to classify samples of the children from different cultures using our classifier and note the accuracy that can be achieved. In [8] similar validation was performed on the condition that was examined in that study and, even though validating microbiome samples came from a different culture, the model's classification accuracy was preserved.

We expect the accuracy and resolution of the sequencing techniques to improve over time. These improvements would result in improved datasets for this type of study. The availability of a dataset with reliable relative abundances beyond the genus-level could contribute to making the accuracy of the classifying model even greater. Our model could be further strengthened by having a larger number of samples along with additional metadata variables. In our future work, we plan to produce adequate additional samples and metadata from children of Bosnia and Herzegovina. We plan to use that data to validate the classifier developed in this study.

## 5  Conclusions

In this study, conducted on microbiome of children of ASD and neurotypical children from Ecuador, we have observed difference in gut microbiome structure that followed most the patterns observed in previous studies. Most notable changes in ASD group compared to Control group were: (i) the decrease of Firmicutes/Bacteroidetes ratio; (ii) an increase in relative abundances of *Faecalibacterium, Ruminococcus, Sutterella, Desulfovibrio* and *Bilophila*; and (iii) a decrease in relative abundance of *Prevotella* and *Lactobacillus*.

The increase in ASD awareness, recognition, diagnosis, and treatment has received substantial attention over previous decades. The ASD global prevalence has been estimated at over 1% [25]. Early diagnosis of ASD is connected to a better disorder outcome. Hence, the goal of this study was to develop a classification model, based on gut microbiome data of ASD and neurotypical children, that could contribute to the early diagnosis of ASD. Our classification model predicted, with high accuracy (over 80%), if a gut microbiome sample came from a child with ASD or not.

## References

1. Ai, L., Tian, H., Chen, Z., Chen, H., Xu, J., Fang, J.Y.: Systematic evaluation of supervised classifiers for fecal microbiota-based prediction of colorectal cancer. Oncotarget **8**, 9546–9556 (2017)
2. Ammaral, D.G.: The promise and the pitfalls of autism. In: Shaw, C.A., Sheth, S., Li, D., Tomljenovic, L.: Etiology of Autism Spectrum Disorders: Genes, Environment, or Both? OA Autism, vol. 2, no. 2, p. 11 (2014)

3. Andreo-Martínez, P., García-Martínez, N., Sánchez-Samper, E.P., Martínez-González, A.E.: An approach to gut microbiota profile in children with autism spectrum disorder. Environ. Microbiol. Rep. **12**, 115–135 (2020)
4. Asgari, E., Garakani, K., McHardy, A.C., Mofrad, M.R.K.: MicroPheno: predicting environments and host phenotypes from 16S rRNA gene sequencing using a k-mer based representation of shallow sub-samples. Bioinformatics **34**(13), i32–i34 (2018)
5. Bolyen, E., et al.: Reproducible, interactive, scalable and extensible microbiome data science using QIIME 2. Nat. Biotechnol. **37**(8), 852–7 (2019)
6. Feng, Z., et al.: A human stool-derived Bilophila wadsworthia strain caused systemic inflammation in specific-pathogen-free mice. Gut Pathog. **9**, 59 (2017)
7. Guang, S., et al.: Synaptopathology involved in autism spectrum disorder. Front. Cell. Neurosci. **21**, 12–470 (2020)
8. Hasic Telalovic, J., Music, A.: Using data science for medical decision making case: role of gut microbiome in multiple sclerosis. BMC Medical Informatics and Decision Making (in print)
9. Ho, L., et al.: Gut microbiota changes in children with autism spectrum disorder: a systematic review. Gut Pathog. **12**, 6 (2020)
10. Iadanza, E., Fabbri, R., Bašić-ČiČak, D., Amedei, A., Telalovic, J.H.: Gut microbiota and artificial intelligence approaches: a scoping review. Health Technol. **10**(6), 1343–1358 (2020). https://doi.org/10.1007/s12553-020-00486-7
11. Kang, D.W., et al.: Long-term benefit of microbiota transfer therapy on autism symptoms and gut microbiota. Sci. Rep. **9**(1), 5821 (2019)
12. Louppe, G.: Understanding random forests. Cornell University Library. arXiv:1407.7502 (2014)
13. McElhanon, B.O., McCracken, C., Karpen, S., Sharp, W.G.: Gastrointestinal symptoms in autism spectrum disorder: a meta-analysis. Pediatrics **133**, 872–83 (2014)
14. McMurdie, P.J., Holmes, S.: phyloseq: an R package for reproducible interactive analysis and graphics of microbiome census data. Plos One **8**(4), e61217 (2013)
15. microbiome R package. Homepage. http://microbiome.github.io. Accessed 20 Sep 2020
16. Modabbernia, A., Velthorst, E., Reichenberg, A.: Environmental risk factors for autism: an evidence-based review of systematic reviews and meta-analyses. Mol. Autism **8**, 13 (2017)
17. Neale, B.M., et al.: Patterns and rates of exonic de novo mutations in autism spectrum disorders. Nature **485**(7397), 242–245 (2012)
18. Pasolli, E., Truong, D.T., Malik, F., Waldron, L., Segata, N.: Machine learning meta-analysis of large metagenomic datasets: tools and biological insights. PLoS Comput. Biol. **12**, e1004977 (2016)
19. Peregosa, F., et al.: Scikit-learn: machine learning in python. J. Mach. Learn. Res. **12**, 2825–2830 (2011)
20. R Core Team. R: A language and environment for statistical computing. R Foundation for Statistical Computing, Vienna, Austria. Homepage (2020). https://www.R-project.org/. Accessed 20 Sep 2020
21. Reiman, D., Metwally, A.A., Yang, D.: PopPhy-CNN: a phylogenetic tree embedded architecture for convolution neural networks for metagenomic data. bioRxiv, p. 257931 (2018)
22. Rivière, A., Selak, M., Lantin, D., Leroy, F., De Vuyst, L.: Bifidobacteria and butyrate-producing colon bacteria: importance and strategies for their stimulation in the human gut. Front. Microbiol. **7**, 979 (2016)
23. Sandin, S., Lichtenstein, P., Kuja-Halkola, R., Hultman, C., Larrson, H., Recihenberg, A.: The heritability of autism spectrum disorder. JAMA **318**(12), 1182–1184 (2017)
24. SILVA: high quality rRNS databases, Homepage. https://www.arb-silva.delast. Accessed 20 Sep 2020

25. Styles, M., et al.: Risk factors, diagnosis, prognosis and treatment of autism. Front. Biosci. **25**, 1682–1717 (2020)
26. Towle, P.O., Patrick, P.A., Ridgard, T., Pham, S., Marrus, J.: Is earlier better? The relationship between age when starting early intervention and outcomes for children with autism spectrum disorder: a selective review. Autism Res. Treat. **2020**, 7605876 (2020)
27. Vuong, H.E., Hsiao, E.Y.: Emerging roles for the gut microbiome in autism spectrum disorder. Biol. Psychiat. **81**(5), 411–423 (2017)
28. Wang, L., Christophersen, C.T., Sorich, M.J., Gerber, J.P., Angley, M.T., Conlon, M.A.: Increased abundance of Sutterella spp. and Ruminococcus torques in feces of children with autism spectrum disorder. Mol. Autism **4**, 42 (2013)
29. Young, H., Oreve, M.J., Speranza, M.: Clinical characteristics and problems diagnosing autism spectrum disorder in girls. Arch. Pediatr. **25**(6), 399–403 (2018)
30. Zhou, Y.H., Gallins, P.: A review and tutorial of machine learning methods for microbiome host trait prediction. Front. Genet. **10**, 579 (2019)

# Industrial Applications

# Deep Ensemble Approach for RUL Estimation of Aircraft Engines

Koceila Abid[1,2(✉)] , Moamar Sayed-Mouchaweh[2], and Laurence Cornez[3]

[1] CEA Tech Hauts de France, 165 Avenue de Bretagne, 59000 Lille, France
[2] IMT Lille Douai, 764 Bd Lahure, 59500 Douai, France
moamar.sayed-mouchaweh@imt-lille-douai.fr
[3] CEA, LIST, 91191 Gif-sur-Yvette, France
laurence.cornez@cea.fr

**Abstract.** Remaining useful life estimation (RUL) is the remaining time until the system failure. Predicting RUL help to schedule the maintenance actions in advance which can improve the reliability and availability of industrial systems while reducing the downtime and maintenance cost. In this paper, a deep ensemble approach for RUL estimation is developed, where the RUL is predicted with two different models: convolutional neural network which is suitable for achieving high level automatic features extraction, and long short term memory is able to capture the temporal information in time series data. The predicted RULs by each model are then aggregated using a weighted mean fusion. The proposed approach is validated using degradation data generated from aircraft engines (C-MAPSS dataset), it can improve the reliability of prediction as well as the accuracy, where it showed promising performance results comparing with the related works in the state of the art.

**Keywords:** Ensemble learning · Convolutional neural network · Long short term memory · Remaining useful life

## 1 Introduction

Prognostics and Health Management (PHM) is a condition based maintenance strategy that is predictive in nature: it aims to determine how long from now a fault may happen in a system given the current operating conditions [6]. PHM strategy is based on four main steps [11], including data acquisition step (data collection and feature extraction), diagnostics step (fault detection, isolation, and identification), prognostics step (remaining useful life estimation), and finally the health management step (actions scheduling for maintenance and logistics). Prognostics is one of the main steps for achieving PHM strategy, it aims to estimate the Remaining Useful Life (RUL) before failure. The RUL is estimated using four global approaches: Reliability, similarity, model based, and data-driven based approaches, where data-driven based approaches has proven

---

Supported by the European Union, European Regional Development Fund.

J. Hasic Telalovic and M. Kantardzic (Eds.): MeFDATA 2020, CCIS 1343, pp. 95–109, 2021.
https://doi.org/10.1007/978-3-030-72805-2_7

its effectiveness and shown a suitable trade-off in terms of precision, applicability, cost, and interpretability [1].

Data-driven approaches require some or several run-to-failure data in order to model the evolution of the degradation and predict the RUL. When there is a lack of available a priori degradation sequences, it is efficient to apply the indirect way for RUL estimation. In this case, a Health Indicator (HI) that best represent the degradation evolution is computed and extrapolated in order to predict its evolution until the failure, while the RUL is deduced as the difference between present time and failure time. This extrapolation can be achieved using adaptive models such as linear model, exponential model, or quadratic regression model [2,3]. On the other hand, when an important number of a priori sequences are collected and available, employing direct RUL estimation is more suitable for an efficient estimation of the RUL. Direct RUL estimation has several advantages: firstly, it is not necessary to understand the operation of the system and its different operating modes, recent data-driven techniques applied for the direct RUL estimation (e.g., CNN and LSTM) can handle the variation of the operating modes in complex dynamic systems. Secondly, it is not necessary to construct or select a suitable health indicator for extrapolation, the direct estimation way can map the raw collected data from sensors directly to the RUL. Finally, it is not necessary to predefine a failure threshold for the RUL estimation, knowing that defining a failure threshold is challenging and may require some domain expertise (Fig. 1).

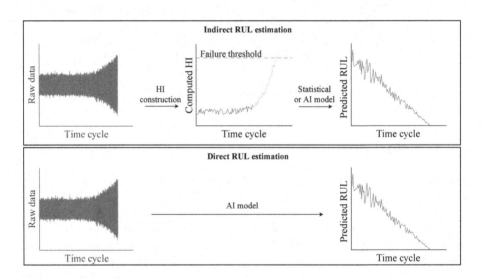

**Fig. 1.** Indirect and direct RUL estimation ways

## 1.1 State of the Art and Related Works

This paper focuses on the case where several a priori run-to-failure sequences are available. Hence, employing direct RUL estimation is suitable for an efficient estimation of the RUL. The paper deals with the RUL estimation of an aircraft engine, where the benchmark data are generated by the National Aeronautics and Space Administration (NASA) and named C-MAPSS dataset [18]. Accordingly, the direct RUL estimation techniques applied for the C-MAPSS dataset are investigated. Machine learning techniques can be applied for this task such as Support Vector Machine (SVM), which is applied to the C-MAPSS problem using the direct RUL estimation [17]. A basic Artificial Neural Network (ANN) named Multi Layer Perceptron (MLP) is used in [16], where the features are extracted in a time window using principal component analysis for RUL estimation. Another ANN architecture is employed in [23] named Extreme Learning Machine (ELM) which has only one hidden layer and is faster in the training phase than the traditional ANN. Recently deeper architectures of the ANN are applied for RUL estimation named Deep Learning model. Deep learning models are more and more used for machine health prognostics and have proven their effectiveness for RUL estimation. The most applied deep learning methods for fault prognostics are Convolutional Neural Network (CNN) and Long Short Term Memory (LSTM). CNN has a deep architecture which makes it suitable for achieving automatic features extraction without computing the features manually. In [5], CNN is applied for direct RUL estimation of data generated from a turbofan engine. The raw data collected from different sensors are used as input whereas the true (actual) RUL is used as target. A deeper CNN architecture is proposed in [15] for RUL estimation, the proposed architecture is deep because it stacks five convolution layers in order to capture the representative information from raw input data. Deep architectures allow learning high level representations of the raw input data, which enhance the prediction performance. LSTM is a recurrent neural network used to learn the long term dependencies between data points. Its architecture enables it to remember information for long periods of time. A direct RUL estimation method is developed in [24], it is based on a simple LSTM which is a network with only one LSTM layer. In [10], the RUL is predicted by using a deeper LSTM architecture which stacks two LSTM layers in order to capture the underlying patterns in the sequential data. Another variant of LSTM named Bi-directional LSTM (BLSTM) is employed in [21], BLSTM is a variant of LSTM which has an architecture that can learn the dependencies of sensor data in both forward and backward direction. Hybrid CNN-LSTM model has also shown high performance for RUL prediction using C-MAPSS data. In [4], an approach based on a hybrid CNN-LSTM model is developed. The hybrid CNN-LSTM aims to extract the features from raw data using CNN, while these features are fed into LSTM for RUL prediction.

## 1.2 Our Approach

A data-driven approach for direct RUL estimation is proposed based on deep ensemble method. This approach combines the decisions of CNN and LSTM

models, which are the most efficient methods for RUL estimation when several
a priori run-to-failure sequences are available. Since the reliability of prediction
is necessary for safety critical systems (e.g., aircraft engine), the proposed deep
ensemble approach can enhance the RUL prediction reliability by fusing the
prediction of two different models (CNN and LSTM). In addition, the proposed
approach has also the ability to predict the RUL of complex dynamic system
(system with several operating modes). Moreover, the proposed deep ensemble
method for direct RUL estimation is efficient to improve the accuracy of RUL
prediction.

The paper is organized as follows: Sect. 2 presents the proposed approach for
RUL estimation. The experimental data used for validation of the proposed deep
ensemble approach are presented in Sect. 3. The obtained results are discussed
in Sect. 4. Finally, Sect. 5 ends the paper with concluding remarks.

## 2   Proposed Approach

The proposed approach is illustrated in Fig. 2 including two phases: an offline
phase for tuning and training the models and an online phase for estimating
the RUL from new incoming observations. In the offline phase (training phase),
the raw data are first preprocessed by selecting the significant input data (sensors), normalizing, segmenting the input data into windows, and setting the
RUL targets (labels). After data preprocessing, different CNN and LSTM models are trained in order to select the optimal hyperparameters by using k-fold
cross-validation. After that, the fusion weights are computed according to the
performance of validation data. In the online phase (testing phase), the data are
preprocessed as in the offline phase. Then, the RULs are predicted using CNN
and LSTM model, while the predicted RULs by each model are then merged
using a weighted mean in order to obtain the final RUL. In the next, CNN and
LSTM models are described, and the decision fusion step is presented.

### 2.1   Data Preprocessing

When the data collected from sensors are acquired, they are first preprocessed
before starting the RUL estimation. The collected raw data from sensors are in
different ranges, this may lead to unequal weight computation in the deep neural
network (CNN and LSTM). Hence, the data are first normalized using z-score
normalization, which is computed as follow:

$$Norm\,(x_s) = \frac{x_s - \mu_s}{\sigma_s} \qquad (1)$$

where $Norm\,(x_s)$ represents the normalized values, $x_s$ are the values of the
sensor $s$, while $\mu_s$ represents the mean and $\sigma_s$ the standards deviation of each
sensor $s$.

The normalized input data are then segmented using a sliding time window,
where the multiple sensors measurement with length $N_f$ are represented as different features. A sliding time window with a fixed length $N_{tw}$ is employed for

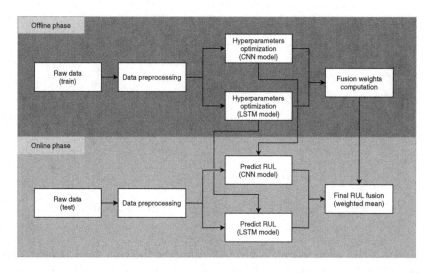

**Fig. 2.** Proposed approach

the segmentation of the consecutive data points, hence, a two dimensional input is obtained for each time cycle ($N_{tw} \times N_f$). In the training phase, this input is fed to the deep learning models while the target of the models is the true RUL.

## 2.2 Convolutional Neural Network Model

Convolutional neural network is developed mainly for computer vision by LeCun [13], it is efficient for automatic feature extraction. The adopted CNN has 1 dimension (1D-CNN), which can handle time series signals. CNN consists of several convolution layers for features extraction, the convolution layer operation is represented as follow:

$$f = \phi(U * k + b) \tag{2}$$

$k$ represents the convolution filter, $U$ is the input data, and $*$ is the convolution operator, while b is the bias term and $\phi$ represents a nonlinear activation function. $f$ is the obtained features map which represents the learned features by sliding the multiple filters on the input data. Deep CNN architecture has proven its efficiency for RUL estimation [15]. More the network is deep more the model can learn high level representation of features. In the proposed CNN architecture, three convolution layers are stacked for efficient features extraction. When the features map is obtained, it is flattened into 1-dimensional shape, and fed to a fully connected layer for RUL prediction as illustrated in Fig. 3.

## 2.3 Long Short Term Memory Model

Long short term memory is an advanced type of recurrent neural network [9], which has been successfully applied for speech recognition and natural language

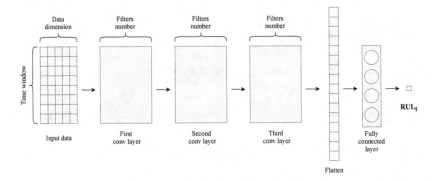

**Fig. 3.** Proposed CNN architecture for RUL estimation

processing. It can address the problem of capturing the long term memory. The LSTM architecture is composed of several connected layers, where each layer contains several connected LSTM cell. The proposed LSTM architecture is presented in Fig. 4, it is a many to one architecture where three LSTM layers are stacked in order to discover the underlying patterns embedded in time series. $x_1, x_2, x_3..., x_t$ are the input data points ($x_1$ is the features vector of the first time index in the time window while $x_t$ is the features vector of the last index of the time window and $RUL_t$ is the predicted RUL (output) of the time window (input).

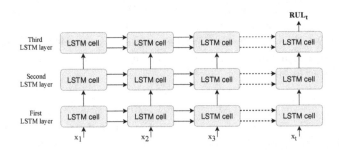

**Fig. 4.** Proposed LSTM architecture for RUL estimation

## 2.4    RUL Fusion

In the online phase, the predicted RULs using the previously described CNN and LSTM models are then aggregated using the weighted mean as illustrated in Fig. 5. Weighted mean is applied in [22] for aggregating the predicted RUL with different time windows. However, the weights are computed according to the training errors, this may increase the weight of overfitted models (when the training error is small whereas the test error is high). For this reason, in our proposed ensemble approach the weights are computed according to a validation error. This is done by using 90% of training sequences to train the models and the remaining 10% of data are used to estimate the validation error. The merged RUL is computed using the following equations:

$$W_k = \frac{\frac{1}{ErrVal_k}}{\sum_{k=1}^{n_k} \frac{1}{ErrVal_k}} \tag{3}$$

$$rul(t) = \sum_{k=1}^{n_k} W_k.rul(t)_k \tag{4}$$

Where $rul(t)$ is the final RUL estimated at each time cycle $t$, $rul(t)_k$ is the RUL estimated by the model $k$ at each time cycle, $n_k$ is the number of models ($n_k = 2$) and $W_k$ is the corresponding weight to each model. $ErrVal_k$ represents the validation RMSE errors for each model $k$.

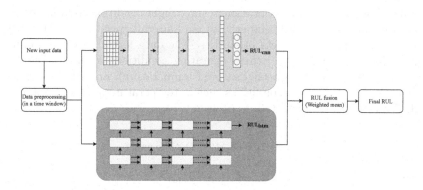

**Fig. 5.** Illustration of RUL fusion for a new input data

A Dropout is applied after the last layer of CNN and LSTM [20], which is a powerful regularization technique that randomly discards a subset of neurons and their connection during training, it is applied to reduce data overfitting when training deep learning models in order to enhance the model generalization. The dropout probability is set to 0.5 which is an optimal probability value for a wide range of networks and tasks [20]. This proposed ensemble approach will improve the accuracy as well as the reliability of RUL prediction for critical systems.

## 3   Experimental Data

The performance of the proposed approach is validated using the benchmark dataset named Commercial Modular Aero-Propulsion System Simulation (C-MAPSS) dataset. The dataset represents the damage propagation of the aircraft gas turbine engines. C-MAPSS dataset is generated by the NASA [18], this

Table 1. C-MAPSS sub-datasets

| Sub-datasets | FD001 | FD002 | FD003 | FD004 |
|---|---|---|---|---|
| Training sequences | 100 | 260 | 100 | 249 |
| Testing sequences | 100 | 259 | 100 | 248 |
| Operating conditions | 1 | 6 | 1 | 6 |
| Fault conditions | 1 | 2 | 2 | 2 |

data has been widely used to compare RUL prediction methods in the literature [4,5,15,17,23].

This dataset can be divided into four cases (or sub-datasets), where each case includes several run-to-failure sequences (or trajectories) for training and for testing. The first and third sub-datasets (FD001 and FD003) are generated under one operating condition, FD001 includes one type of fault, while FD003 includes two fault types. The second and fourth cases (FD002 and FD004) are generated under 6 operating conditions (variation of three flight conditions parameters: aircraft altitude, environmental temperature, and aircraft speed). Table 1 summarize the C-MAPSS sub-datasets properties. The data format contains columns about: unit index, time cycle, three operating conditions or flight conditions (altitude, mach number or speed, sea level temperature), and 21 sensors measurements about the system conditions such as temperature, pressure, rotational speed. Data format is illustrated in Table 2.

Table 2. C-MAPSS Data format

| Column n# | 1 | 2 | 3–5 | 6–26 |
|---|---|---|---|---|
| Information | Unit index | Time cycle index | Operating conditions | Sensor values |

The training data contains run-to-failure sequences from healthy to failure, while the test data contains sequences that stop at some time before failure. The goal is to estimate the RUL of the test data until failure. Then, the predicted RUL should be evaluated for all the engine units according to the true RUL (actual RUL) which is provided in the datasets. Two evaluation criteria are employed: Root Mean Squared Error (RMSE) and the value of a scoring function defined in [18]. The scoring function penalizes more the overestimated RUL. The two metrics are calculated as follows:

$$RMSE = \sqrt{\frac{1}{n_u} \sum_{u=1}^{n_u} (rul_{pred}(u) - rul_{true}(u))^2} \tag{5}$$

$$Score = \begin{cases} \sum_{u=1}^{n_u} \left( e^{\frac{-d(u)}{13}} - 1 \right) & \text{for} \quad d(u) < 0 \\ \sum_{u=1}^{n_u} \left( e^{\frac{d(u)}{10}} - 1 \right) & \text{for} \quad d(u) \geq 0 \end{cases} \tag{6}$$

where $n_u$ is the total number of engine units in the test data, $u$ is the engine units index, $rul_{true}$ represents the actual RUL, and $rul_{pred}$ represents the predicted RUL. $d(u)$ is the difference between the predicted and true RUL $(rul_{pred}(u) - rul_{true}(u))$.

## 4   Results and Discussion

In the training phase, the data are first preprocessed before modeling. Firstly, the data collected by sensors n# 2, 3, 4, 7, 8, 9, 11, 12, 13, 14, 15, 17, 20, and 21 are selected ($N_f = 14$), because the values of the discarded sensors remain unchanged during operation. In addition to the selected sensors, the operating conditions (flight conditions) measurement are also selected for the cases FD002 and FD004 due to the variation of the operating modes as illustrated in Fig. 6.

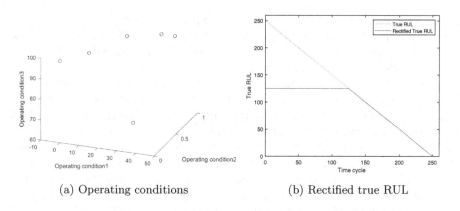

(a) Operating conditions                    (b) Rectified true RUL

**Fig. 6.** Illustration of Operating modes and rectified true RUL

In Fig. 6a, operating condition 1 refers to the altitude from sea level ($10^3$ ft), operating condition 2 represents the mach number (a ratio of flow velocity to the speed of sound), and operating condition 3 is the sea-level temperature (°F). For FD001 and FD003 there is only 1 mode of operating condition represented by the red circle, where for FD002 and FD004 there are 6 different modes of operating conditions represented by red and blue circles. This variation of operating conditions modes results in a variation of the sensor values, which may hide the observation of the system degradation with time. Figure 7a represents the values of sensor number #2 in the sub-datasets FD001, the degradation can be observed on the sensor values which is monotonic and increase with time because there is only one mode of operating conditions. However, in Fig. 7b which represents the values of sensor number #2 in the sub-datasets FD002, the degradation evolution over time cannot be observed due to the variation of the operating modes (change in operating conditions), this operating modes variation may increase the difficulty of RUL estimation.

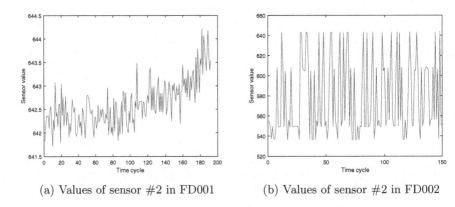

(a) Values of sensor #2 in FD001          (b) Values of sensor #2 in FD002

**Fig. 7.** Sensor values under one and six operating conditions

In general, the target RUL (true RUL) of a training set sequence should be inversely proportional to the time cycle. For the C-MAPSS datasets, a piece-wise linear function is proposed by [8] to rectify the training labels. When the training cycle is smaller than a predefined threshold, the true RUL is constant, and the system is considered healthy in this case. Then, when the training cycle is greater than the threshold, the target RUL starts decreasing (as illustrated in Fig. 6b). According to the literature the threshold is set to 125 [4,14,15]. This rectification is set because the RUL should not decrease at the beginning of the degradation since the system is considered always under healthy conditions. Also, this rectification will prevent overestimating the RUL prediction (the evaluation score will be larger). Moreover, the principal reason for setting this rectification is to allow a suitable comparison in the same conditions with the related works.

The input of the deep learning models should be in a window, the length of the time window ($N_{tw}$) should be large enough in order to include the maximum of information. On the other side, the minimum running cycle length of the sequences in the test data is 31 cycles, therefore the window length should be less than 31 cycles. Time window of length 15 and 30 cycles are compared for RUL prediction accuracy in [4], where the time window with a length of 30 cycles showed better performance, hence for this work, the selected sensor values are normalized using z-score normalization and then segmented using a time window with the length of $N_{tw} = 30$ time cycles.

For training, the RMSE is used as cost function, back-propagation learning is utilized for the updates of the weights in the network using mini-batches. Adam optimizer algorithm [12] is used for training the two models. A kernel size of 10 is adopted which is applied in [15] and showed a good performance on C-MAPSS data. An early stopping criterion is applied in order to stop the training of the network, the training is stopped when the validation is not decreasing for 10 epochs (iterations). For choosing the best hyperparameters (mini-batch size $Bs$, learning rate $Lr$, filters number $F$ for CNN, and hidden unit $L$ in LSTM) for each model, 5k-fold cross validation is applied where each model is trained 5 times with 80% of the training sequences and evaluated with the remaining 20% of the training sequences, the results are shown in Table 3. Since neural networks learning is non convex, the models with the selected hyperparameters are trained 10 times, and the best models according to the validation are selected, in this step, the training is split 90% for training and 10% for validation.

**Table 3.** Hyperparameters selection for each subdataset

| Subdataset | CNN hyperparameters | LSTM hyperparameters |
|------------|---------------------|----------------------|
| FD001 | Bs(512)F(64,64,64)Lr(0.0001) | Bs(512)L(32,32,32)Lr(0.01) |
| FD002 | Bs(128)F(16,16,16)Lr(0.0001) | Bs(128)F(16,16,16)Lr(0.001) |
| FD003 | Bs(512)F(32,32,32)Lr(0.0001) | Bs(512)L(32,32,32)Lr(0.01) |
| FD004 | Bs(128)F(16,16,16)Lr(0.0001) | Bs(128)F(16,16,16)Lr(0.001) |

The efficiency of the ensemble based method is shown in Table 4, it presents the prediction errors RMSE and score using CNN, LSTM, and the proposed deep ensemble model for each subdataset. It can be observed that the separated CNN and LSTM show a good RUL accuracy (RMSE and Score), while the deep ensemble model reveal better accuracy than each separated model. RMSE and score are slightly high for FD002 and FD004, this is because the RUL prediction is difficult due to the variation in the operating modes.

**Table 4.** Evaluation of the CNN, LSTM, and ensemble model

| Model | FD001 | | FD002 | | FD003 | | FD004 | |
|-------|-------|-------|-------|-------|-------|-------|-------|-------|
| | RMSE | Score | RMSE | Score | RMSE | Score | RMSE | Score |
| CNN | 11.48 | 176.94 | 15.46 | 1287.56 | 12.11 | 235.52 | 23.20 | 6572.25 |
| LSTM | 12.02 | 249.50 | 14.57 | 982.51 | 12.39 | 238.02 | 17.03 | 1151.78 |
| **Deep ensemble** | **10.74** | **176.36** | **14.23** | **984.34** | **11.48** | **206.53** | **18.05** | **1478.70** |

The RUL prediction result of the testing engine units in FD001 are presented in Fig. 8a. The testing units are sorted according to the true RULs (test labels) from small to large (for better visualization). It can be observed that the RUL prediction is more accurate when the engine is near to failure (true RUL is small), this is because the degradation is in a significant level and information about the degradation can be seen in the features data, hence, our method is able to capture this degradation level and become more accurate when approaching the failure. Figure 8b is a zoom of the Fig. 8a for the test unit #60, it can be seen in that the predicted RUL using each separated model CNN and LSTM is far from the true RUL. However, the proposed deep ensemble method shows a better RUL prediction near to the true RUL, this is performed thanks to

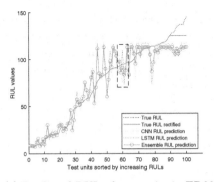 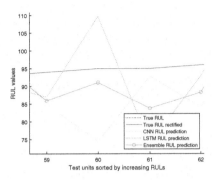

(a) Predicted RUL of test units in FD001     (b) Predicted RUL for test unit #60

**Fig. 8.** Predicted RUL for the test unit #60 in FD001

the proposed fusion method where the predicted RULs using each model are aggregated by using the weighted mean. The predicted RUL for the test units in each subdataset are presented in Fig. 9.

**Table 5.** Performance comparison with the related works on C-MAPSS dataset

| Method | Year | FD001 | | FD002 | | FD003 | | FD004 | |
|---|---|---|---|---|---|---|---|---|---|
| | | RMSE | Score | RMSE | Score | RMSE | Score | RMSE | Score |
| SVM [17] | 2013 | 29.82 | – | – | – | – | – | – | – |
| MLP [16] | 2016 | 15.15 | – | - | – | – | – | – | – |
| CNN [5] | 2016 | 18.44 | 1286.70 | 30.29 | 13570 | 19.81 | 1596.20 | 29.15 | 7886.4 |
| LSTM [24] | 2017 | 16.14 | 338 | 24.49 | 4450 | 16.18 | 852 | 28.17 | 5550 |
| LSTM [10] | 2018 | 16.73 | 388.68 | 29.43 | 10654 | 18.06 | 822.19 | 28.39 | 6370 |
| ELM [23] | 2018 | 13.78 | 267.31 | – | – | – | – | – | – |
| BLSTM [21] | 2018 | 13.65 | 295 | 23.18 | 4130 | 13.74 | 317 | 24.86 | 5430 |
| CNN [15] | 2018 | 12.61 | 273.7 | 22.36 | 10412 | 12.64 | 284.1 | 23.31 | 12466 |
| Stacking ensemble [19] | 2019 | 16.67 | – | 25.57 | – | 18.44 | – | 26.76 | – |
| BHLSTM [7] | 2019 | | 376.64 | | | | 1422 | | |
| Hybrid CNN-LSTM [4] | 2019 | 13.017 | 245 | 15.24 | 1282.42 | 12.22 | 287.72 | 18.15 | 1527.42 |
| Hybrid CNN-BLSTM [22] | 2020 | 12.66 | 304.29 | – | – | – | – | – | – |
| MS-DCNN [14] | 2020 | 11.44 | 196.22 | 19.35 | 3747 | 11.67 | 241.89 | 22.22 | 4844 |
| **Ensemble CNN-LSTM (proposed approach)** | **2020** | **10.74** | **176.36** | **14.23** | **984.34** | **11.48** | **206.53** | **18.05** | **1478.70** |

The performance of our proposed approach is compared with the related works applied to the C-MAPSS dataset. Table 5 summaries the latest research results sorted in the ascending order of publication year. From the table, it can be seen that recently deep learning models are widely applied and gives better results compared to the traditional machine learning models such as SVM

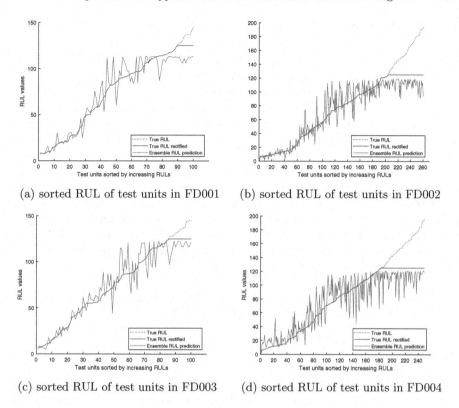

(a) sorted RUL of test units in FD001    (b) sorted RUL of test units in FD002

(c) sorted RUL of test units in FD003    (d) sorted RUL of test units in FD004

**Fig. 9.** RUL prediction for the test units for all the subdatasets

or MLP. It can be also observed that the RMSE and scoring errors are lower for FD001 and FD003 than FD002 and FD004, this is due to the viability of the operating conditions modes in the subdatasets FD002 and FD004. Our proposed approach outperforms the other related works in terms of accuracy of prediction (RMSE and Score are low). Our proposed approach has shown better performance than other ensemble approach such as MS-DCNN [14], where they develop an approach that uses an ensemble of CNN models with different time window lengths. The proposed approach also outperforms the hybrid models such as hybrid CNN-LSTM [4] and hybrid CNN-BLSTM [22] and is more reliable than these models because the RULs are predicted with two different models, while the predicted RULs are merged in order to obtain the final RUL. This can confirm the suitability of exploiting the advantage of each model (CNN and LSTM) by fusing their decision (RUL estimation), it can improve the accuracy of prediction as well as the reliability.

## 5   Conclusion

In this paper, a deep ensemble approach is proposed which exploits the diversity of two different deep learning models named CNN and LSTM. CNN architecture can extract relevant information by applying several convolution filters on the raw data, while LSTM is able to capture the temporal information in time series data. The proposed deep ensemble approach for RUL estimation is validated using the well known C-MAPSS dataset and has achieved promising performance compared with the state-of-the-art results.

Since the reliability of prediction is necessary for safety critical systems such as aircraft engines, the proposed deep ensemble approach can enhance the RUL prediction reliability by fusing the prediction of two different models. The proposed approach has also shown its ability to capture the variability of the operating modes (flight conditions). Finally, the proposed deep ensemble method for direct RUL estimation has proven its efficiency as shown by the results since it can improve the accuracy of RUL prediction.

**Acknowledgment.** This paper is the result of the research work supported by the European Union, European Regional Development Fund.

## References

1. Abid, K., Sayed Mouchaweh, M., Cornez, L.: Fault prognostics for the predictive maintenance of wind turbines: state of the art. In: Monreale, A., et al. (eds.) ECML PKDD 2018. CCIS, vol. 967, pp. 113–125. Springer, Cham (2019). https://doi.org/10.1007/978-3-030-14880-5_10
2. Abid, K., Sayed-Mouchaweh, M., Cornez, L.: Adaptive machine learning approach for fault prognostics based on normal conditions-application to shaft bearings of wind turbine. In: Proceedings of the Annual Conference of the PHM Society, vol. 11 (2019)
3. Ahmad, W., Khan, S.A., Kim, J.M.: A hybrid prognostics technique for rolling element bearings using adaptive predictive models. IEEE Trans. Industr. Electron. **65**(2), 1577–1584 (2017)
4. Al-Dulaimi, A., Zabihi, S., Asif, A., Mohammadi, A.: A multimodal and hybrid deep neural network model for remaining useful life estimation. Comput. Ind. **108**, 186–196 (2019)
5. Sateesh Babu, G., Zhao, P., Li, X.-L.: Deep convolutional neural network based regression approach for estimation of remaining useful life. In: Navathe, S.B., Wu, W., Shekhar, S., Du, X., Wang, X.S., Xiong, H. (eds.) DASFAA 2016. LNCS, vol. 9642, pp. 214–228. Springer, Cham (2016). https://doi.org/10.1007/978-3-319-32025-0_14
6. Das, S., Hall, R., Patel, A., McNamara, S., Todd, J.: An open architecture for enabling CBM/PHM capabilities in ground vehicles. In: 2012 IEEE Conference on Prognostics and Health Management, pp. 1–8. IEEE (2012)
7. Elsheikh, A., Yacout, S., Ouali, M.S.: Bidirectional handshaking LSTM for remaining useful life prediction. Neurocomputing **323**, 148–156 (2019)

8. Heimes, F.O.: Recurrent neural networks for remaining useful life estimation. In: 2008 International Conference on Prognostics and Health Management, pp. 1–6. IEEE (2008)
9. Hochreiter, S., Schmidhuber, J.: Long short-term memory. Neural Comput. **9**(8), 1735–1780 (1997)
10. Hsu, C.S., Jiang, J.R.: Remaining useful life estimation using long short-term memory deep learning. In: 2018 IEEE International Conference on Applied System Invention (ICASI), pp. 58–61. IEEE (2018)
11. Kim, N.-H., An, D., Choi, J.-H.: Prognostics and Health Management of Engineering Systems. Springer, Cham (2017). https://doi.org/10.1007/978-3-319-44742-1
12. Kingma, D.P., Ba, J.: Adam: a method for stochastic optimization. arXiv preprint arXiv:1412.6980 (2014)
13. LeCun, Y., Bengio, Y., et al.: Convolutional networks for images, speech, and time series. In: The Handbook of Brain Theory and Neural Networks, vol. 3361, no. 10 (1995)
14. Li, H., Zhao, W., Zhang, Y., Zio, E.: Remaining useful life prediction using multi-scale deep convolutional neural network. Appl. Soft Comput. **89**, 106113 (2020)
15. Li, X., Ding, Q., Sun, J.Q.: Remaining useful life estimation in prognostics using deep convolution neural networks. Reliab. Eng. Syst. Saf. **172**, 1–11 (2018)
16. Lim, P., Goh, C.K., Tan, K.C.: A time window neural network based framework for remaining useful life estimation. In: 2016 International Joint Conference on Neural Networks (IJCNN), pp. 1746–1753. IEEE (2016)
17. Louen, C., Ding, S., Kandler, C.: A new framework for remaining useful life estimation using support vector machine classifier. In: 2013 Conference on Control and Fault-Tolerant Systems (SysTol), pp. 228–233. IEEE (2013)
18. Saxena, A., Goebel, K., Simon, D., Eklund, N.: Damage propagation modeling for aircraft engine run-to-failure simulation. In: 2008 International Conference on Prognostics and Health Management, pp. 1–9. IEEE (2008)
19. Singh, S.K., Kumar, S., Dwivedi, J.: A novel soft computing method for engine RUL prediction. Multimedia Tools Appl. **78**(4), 4065–4087 (2019). https://doi.org/10.1007/s11042-017-5204-x
20. Srivastava, N., Hinton, G., Krizhevsky, A., Sutskever, I., Salakhutdinov, R.: Dropout: a simple way to prevent neural networks from overfitting. J. Mach. Learn. Res. **15**(1), 1929–1958 (2014)
21. Wang, J., Wen, G., Yang, S., Liu, Y.: Remaining useful life estimation in prognostics using deep bidirectional LSTM neural network. In: 2018 Prognostics and System Health Management Conference (PHM-Chongqing), pp. 1037–1042. IEEE (2018)
22. Xia, T., Song, Y., Zheng, Y., Pan, E., Xi, L.: An ensemble framework based on convolutional bi-directional LSTM with multiple time windows for remaining useful life estimation. Comput. Ind. **115**, 103182 (2020)
23. Zheng, C., et al.: A data-driven approach for remaining useful life prediction of aircraft engines. In: 2018 21st International Conference on Intelligent Transportation Systems (ITSC), pp. 184–189. IEEE (2018)
24. Zheng, S., Ristovski, K., Farahat, A., Gupta, C.: Long short-term memory network for remaining useful life estimation. In: 2017 IEEE International Conference on Prognostics and Health Management (ICPHM), pp. 88–95. IEEE (2017)

# Designing an Indoor Real-Time Location System for Healthcare Facilities

Noemi Falleri[1]([⊠]), Alessio Luschi[2]([⊠]), Roberto Gusinu[3]([⊠]), Filippo Terzaghi[3]([⊠]), and Ernesto Iadanza[2]([⊠])

[1] Bachelor's Degree in Electronic and Telecommunications,
University of Florence, Firenze, Italy
`noemi.falleri@stud.unifi.it`
[2] Department of Information Engineering, University of Florence, Firenze, Italy
`{alessio.luschi,ernesto.iadanza}@unifi.it`
[3] Azienda Ospedaliero Universitaria Senese, Siena, Italy
`{roberto.gusinu,filippo.terzaghi}@ao-siena.toscana.it`

**Abstract.** This paper performs a feasibility analysis to find the best-fitting solution in terms of quality/price ratio for designing and developing a Real-Time Location System for Indoor Positioning inside healthcare facilities. In particular, an overall comparison of all the available solutions is done, highlighting pros and cons of each technology (WiFi, RFID, WLAN, Ultra-Wide Band, Bluetooth Low Energy, ZigBee, magnetic fields, infrareds, ultrasounds, computer-vision and Pedestrian Data-Reckoning) for accuracy, price, coverage, infrastructure development and installation, and maintenance. A preliminary scope-review is also produced, which summarizes the obtained outcomes in similar studies. In the results section the proposed system is illustrated via flowcharts and block diagrams, both for off-site and on-site scenarios.

**Keywords:** RTLS · IPS · Healthcare · Hospitals

## 1  Introduction

Navigation portable applications have largely grown during the last years, especially because of the huge diffusion of smartphones with inner localization hardware, such as Global Navigation Satellite System (GNSS), wireless antennas (IEEE 802.11 WiFi and IEEE 802.15 Bluetooth) and inertial sensors. However, the majority of these applications works just for outdoor positioning and routing, due to their architecture based upon GPS (Global Positioning System) signals, which have extremely low power, therefore they cannot be received inside a building.

New technologies and algorithms have been developed and adopted during the recent years to face Indoor Positioning System (IPS), such as Radio-frequency Identification (RFID), Wireless Local Area Network (WLAN), Ultra-wide Band (UWB), Bluetooth Low Energy (BLE), ZigBee, magnetic fields, infrared (IR), ultrasounds (US) or computer-vision based systems [1–3].

© Springer Nature Switzerland AG 2021
J. Hasic Telalovic and M. Kantardzic (Eds.): MeFDATA 2020, CCIS 1343, pp. 110–125, 2021.
https://doi.org/10.1007/978-3-030-72805-2_8

Real-time Location Systems (RTLS) is a topic of research which is becoming wider and wider. No universal standard is currently available, due to the peculiarities of each building and to the positioning requirements each application has. In fact, mixed technologies are often implemented depending on the features of the spaces. The application architecture is also very heterogenous depending on the type of users it is developed for. A drive system for robots requires a millimetric precision while a public application for general users, needs less accuracy, but on the other hand, it must be compliant with multiple operative systems and hardware of different brands of smartphones. Moreover, some of these systems only works when additional infrastructures have been installed, which result in extra price and maintenance. Consequentially, the building and field of application affect the system and the algorithms to choose [4].

The scope of this paper is to find the best-fitting solution for developing a RTLS for the hospital of Le Scotte in Siena (Italy). The hospital is made out of 12 pavilions and covers an area of about 208,000 m$^2$ with 800 beds and 8,100 rooms. Each pavilion is located on a topography which is largely hilly, thus the inner paths and alleys which link together different buildings are not on the same constant level throughout the hospital area. In fact, it is very likely to have the first or second basement storey of a building at the same elevation of the ground floor of an adjoining pavilion. This implies a very complex spatial management of the whole premise [5–7]. This spatial unicity together with the heterogeneity of the users of a hospital (patients with limited mobility, visitors, suppliers, technicians) represents a valid opportunity for implementing a navigation system in order to improve the user's experience. The first obstacles which come to mind in applying such technology to a hospital is the limited availability of a consistent WiFi coverage throughout the whole premise together with the presence of a lot of metal devices, some of which may also be moved.

## 2 Methods

An Indoor Position System is a network of devices which allows accurate and real-time indoor people and items localization. Generally, an IPS can be divided in 3 main blocks:

- Inner positioning system module
- Navigation module
- Human-machine interaction (HMI) module.

The first one estimates the user's spatial position, the second one evaluates the available routes between the starting point and the final destination, while the latter improves the user's interaction with the system and gives him/her useful instructions and information [1].

Several technologies are available for implementing the inner positioning system module, which can be categorised in: radio frequencies (WLAN, WiFi, RFID, UWB, ZigBee and Bluetooth), magnetic fields, computer-vision based systems, IR systems, US systems, Pedestrian Dead-Reckoning (PDR). Each of the above listed wireless technology utilises a dedicated positioning model which may change in terms of coverage and precision according to the selected medium (electromagnetic waves, optical waves or mechanical waves). The main criteria to keep in consideration while choosing the appropriate technology are: accuracy (the average Euclidian distance between the actual ground position and the estimated position coordinates), precision, coverage, scalability, medium and infrastructure, robustness, power consumption, price, usability, safety and privacy [1].

A list of all the available technologies follows, highlighting the pros and cons for each one (which are also summarised in Table 1).

1. **WiFi** is a RF technology used in WLAN based upon the standard IEEE 802.11. IPS which uses WiFi infrastructure is often implemented in internal environments with an already existing access point (AP) architecture for data transferring. In order to adopt this technology for a full RTLS, multiple wireless access points (WAP) need to be installed so that the actual coverage can be increased, improving the accuracy of localizing items and people, with smartphones acting as wireless clients. In WiFi RTLS, the accuracy is about 3–5 m, which is a non-acceptable value for the scope of this study, even though new approaches are being developed to improve the precision of the measurement [8–10]. Another disadvantage is the lack of API for indoor localization via WiFi for iOS devices. Apple Inc. Stopped designing and deploying API for signal strength detection via WAP. This results in a practical difficulty in designing a universal system, requiring the installation of iBeacon as a support for Received Signal Strength Indication (RSSI) [11].

2. **RFID** systems are made of tags which contain data that can be recovered with a radiofrequency reader by employing the Received Signal Strength (RSS), the Angle of Arrival (AOA), the Time of Arrival (TOA) or the Time Difference of Arrival (TDOA) for estimating their position. RFID tags can be active, passive or semi-active. Current RTLS based on RFID use passive tags because they do not need any internal power supply: the radio waves emitted by the reader provide sufficient energy to transmit the whole data. This results in a limited reading area of a few meters: the price of the system is kept low, but the number of needed antennas increase. Another problem which need to be faced when adopting passive tagging is whenever the reader is not able to evaluate multiple responses from different tags at the same time (collision), affecting the scalability. Moreover, metals may cause electromagnetic interferences and distortions. Finally, all the above-cited approaches, except RSS, may not be able to accurately estimate the tag position in an indoor environment.

3. **UWB** is a radio technology that can use a very low energy level for short-range, high-bandwidth communications over a large portion of the radio spectrum. They ensure very precise spatial position estimation with deviation of about 0.01 m and a wide signal coverage of about 30 m. The high-bandwidth ensures a high data-transfer speed together with a high robustness, also in a multipath environment. The system uses TOA, AOA, TDOA and RSS for position estimation just like RFID. UWB is the most accurate system but it is very expensive due to its type of tag and infrastructure. Besides, the installation may also be very complex [12].

4. **ZigBee** is an IEEE 802.15.4-based specification for a suite of high-level communication protocols used to create personal area networks with small, low-power digital radios, designed for small scale projects which need wireless connection. Hence, Zigbee is a low-power and close proximity (i.e., personal area) wireless

ad hoc network. The high coverage (between 10 and 100 m) and the low-power working rate is however at the expense of the data rate and precision (about 5 m).

5. **Bluetooth Low Energy** (Bluetooth LE or BLE) is a wireless personal area network technology designed and marketed by the Bluetooth Special Interest Group (Bluetooth SIG). Compared to Classic Bluetooth, BLE is intended to provide considerably reduced power consumption and price while maintaining a similar communication range. Mobile operating systems including iOS, Android, Windows Phone and BlackBerry, as well as macOS, Linux, Windows 8 and Windows 10, natively support Bluetooth Low Energy. Bluetooth beacons are usually used as RF sources to trace the devices' position by implementing proximity detection, RSS fingerprinting or trilateration. Accuracy is of 2–3 m, with a coverage of 10–20 m (which can be widen at the expense of the battery duration).

6. **Magnetic fields** localization systems rely on the interferences caused by structural steel elements of the building to the earth magnetic field, which produce unique magnetic prints. This technique finds numerous advantages, such as no pre-implemented infrastructure, low price and no influences by human bodies or any other kind of barriers. Geomagnetic field ensures high precision despite a low-energy consumption with a subsequent mobile battery saving [13]. However, it may not perform well in a large area [4, 14] and it takes long time to build the initial magnetic map, which also needs updates every time a metal asset or furniture is added or removed to the scene [15].

7. **Vision-based** approach utilises computer-vision algorithm to place an image framed by a smartphone inside a 3D scene, by recognizing key items, shapes or texts. This technology offers scalable systems at a low price, but it affects the accuracy because the device needs to remain in a stable vertical position while targeting key pictures [16].

8. **Infrared positioning systems** are based upon IR receivers which can establish the location of IR transmitters spread throughout the building. IR have a very restricted coverage and require a Line of Sight (LoS) between transmitters and receivers. Furthermore, they are very sensitive to interferences of other IR sources (such as the sun itself) and they also need expensive hardware and maintenance.

9. **Ultrasounds** can also be used to define the position of ultrasonic tags, in addition to radio waves and IR. Unfortunately, US are more sensitive to obstacles then radio waves, and they are obviously also sensitive to sound interferences and to the heat [17].

10. **Pedestrian Dead-Reckoning** (PDR) technique estimates the position of a device by knowing its starting point, direction and travelled distance. Smartphones have one or more built-in inertial sensors (accelerometer, gyroscope) to perform the desired evaluation. Usually, these systems obtain the starting position by using other methodologies and then use the smartphone's accelerometer to get to know when and where the user steps over. Position is evaluated at every step by using the previous known position: in the long run, this brings to an additive error reliant to the sensor's precision (drift-error). On the other hand, these systems do not need any external infrastructure and they are not subject to external interferences.

**Table 1.** Highlighting pros and cons of different RTLS technologies.

|  | Accuracy [m] | Range [m] | Price | Pros | Cons |
|---|---|---|---|---|---|
| Wi-Fi | 1 ÷ 5 | 20 ÷ 30 | Medium | Existing infrastructure in modern buildings | iOS devices need iBeacons; high energy consumption |
| Passive RFID | 2 ÷ 3 | 2 ÷ 3 | Low | Passive tagging | Small coverage; tag collision; electromagnetic interferences |
| UWB | Less than 1 | 20 ÷ 30 | High | High accuracy | Expensive tags and infrastructure; complex installation |
| ZigBee | 3 ÷ 5 | 10 ÷ 100 | High | Mesh topology; wide range | Low precision; long implementation times to reduce price |
| BLE | 2 ÷ 3 | 10 ÷ 20 | Low | Low energy consumption; scalability | Expensive infrastructure and maintenance |
| Magnetic Field | Less than 1 | 1 ÷ 10 | Null | High accuracy; no infrastructure needed; low price | Not suitable for wide area; complex mapping |
| Computer Vision | 1 ÷ 3 | N/A | Null | Low price; no infrastructure needed; scalable | Stability of devices during image acquisition |
| IR | Less than 1 | 1 ÷ 5 | Medium/High | Good precision at room level | LoS; interferences; short range |
| US | Less than 1 | 2 ÷ 10 | Medium/High | Accuracy | LoS; interferences |
| PDR | 1 ÷ 6 | Null | Null | No infrastructure needed; low price; simplicity | Drift-error; recalibration needed |

Some of the described technologies can be already excluded from this study because they do not have properties which well suit the features of a healthcare facility: US and IR systems are too sensitive to interference with very common signal sources for hospitals, and ZigBee is specifically designed for low-scale projects. Moreover, UWB is excluded due to its high price. RFID is often used in hospitals for medical device [18] and patient tracking [19] or sampling recording, but it is not a successful solution for RTLS purposes [2, 20].

The remaining technologies (WiFi, BLE, magnetic field, computer vision and PDR) are analysed in detail by comparing the dedicated literature in Tables 2, 3, 4 and 5.

**Table 2.** RF systems applications for indoor navigation.

| Authors | Method | System | Performance | Test size |
|---------|--------|--------|-------------|-----------|
| Sadowski S. e Spachos P. [21] | ZigBee, BLE and WiFi | Trilateration | Average error: ZigBee: 5.1317 m BLE: 1.1143 m WiFi: 0.5183 m | 5.6 m × 5.9 m |
| Wang X. et al. [22] | WiFi | BiLoc, bi-modal deep learning and fingerprinting | Errore medio TEST 1: 1.5743 m Errore medio TEST 2: 2.5101 m | Test 1: 6 × 9 m$^2$ Test 2: 2.4 × 24 m$^2$ |
| Ibrahim M. et al. [8] | WiFi | WiFi fingerprinting and Fuzzy logic | Average error: 1–2 m Max error: 3–4 m Accuracy: <2 m al 95% | 10 m$^2$ |
| Joseph R. e Sasi S. [23] | WiFi | Fingerprinting | Accuracy: 93% above 20 interactions | – |
| Yu J. et al. [24] | WiFi, PDR | Fingerprinting, KDE and PDR, UKF | Average error: 0.76 m | 43.5 × 11.2 m$^2$ |
| Abdulkarim H.D. et al. [25] | WiFi, PDR | RSS normalization proximity values, EKF integrated PDR (self-calibration extended Kalman filter) | Average error: Non-normalized: 2.05 m Normalized RSS: 1.96 m | 179 m$^2$ |
| Tian Z. et al. [26] | WiFi, Micro Electro-Mechanical Systems (MEMS) | Fingerprinting WiFi, EKF integrated PDR | RMSE: 0.8 m Accuracy: 90% < 1.7 m | TEST 1: 64.6 × 18.5 m$^2$ TEST 2: 81.2x18.5 m$^2$ |
| Cui Y. et al. [10] | WiFi, MEMS | Fingerprinting WiFi and SKF integrated PDR (self-calibration Kalman filter) | Average error: 0.6086 m | 110 × 30 m$^2$ |

**Table 3.** Geomagnetic field systems applications for indoor navigation.

| Authors | Method | System | Performance | Test size |
|---|---|---|---|---|
| Selamat M.H. e Narzullaev A. [27] | WiFi and Magnetic field comparison | Fingerprinting WiFi, fingerprinting for magnetic field | Average error: Magnetic field: cm WiFi: 1–3 m | – |
| Ashraf I. et al. [28] | Magnetic field integrated with smartphones' sensors | Fingerprinting, PDR | Accuracy: Galaxy S8: 50% 0.88 m 75% 1.68 m LG G6: 50% 1.21 m 75% 2.20 m | $85 \times 40$ m$^2$ $50 \times 35$ m$^2$ $30 \times 30$ m$^2$ $50 \times 35$ m$^2$ $90 \times 32$ m$^2$ |
| Chen Y. et al. [29] | Magnetic field | Fingerprinting, Magnetic Field Sorting (MFS) | Average error: [2.13–3.27]m | – |
| Shu Y. et al. [30] | Magnetic field | Magicol | Accuracy: 80%: Office: 4 m Market: 3.5 m Underground parking: 1 m | Office:4000 m$^2$ Market:1900 m$^2$ Underground parking: 3800 m$^2$ |
| Chen L. et al. [13] | Magnetic field | MeshMap | Accuracy: 70% < 2 m 95% < 4 m | – |
| Li P. et al. [31] | Magnetic field | Converging Stepped Magnetofluid Seal (CSMS) by integrating Chemical Shift-resolved Spectroscopic Imaging (CSI) and MFS fingerprinting | Average error: 0.5 m | Laboratorio: 8 m × 20 m Corridoio: 2.4 m 30 m |
| Lee N. et al. [32] | Magnetic field | Accurate Magnetic Indoor Localization (AMID), deep learning | Hallway [1]/Lobby [2] Average error: [1]: 0.76 m /[2]: 2.30 m Accuracy 90%: [1]:1.50 m/[2]: 8.14 m Accuracy 50%: [1]:0.60 m/[2]:0.90 m | Hallway: 15 m 65 m Lobby: 15 m 22 m |
| Bhattarai B. et al. [14] | Magnetic field | Fingerprinting, Deep Recurrent Neural Network (DRNN) based on Long Short-term Memory (LSTM) | Accuracy: 97.2% | Hallway: 100 m × 2.5 m Lab: 7 m × 7 m |
| Ning F.S. et al. [33] | Magnetic field integrated with smartphones' inertial sensors | PDR, magnetic field mapping | Average error: Male: 1 m/Female: 0.6 m Accuracy: 80% < 1 m 50% < 0.64 m | 33 m × 85 m |

**Table 4.** Hybrid systems applications for indoor navigation.

| Authors | Method | System | Performance | Test size |
|---|---|---|---|---|
| Li Y. et al. [34] | WiFi, Magnetic field and inertial sensors | WiFi fingerprinting, magnetic matching (MM), PDR | RMS:<br>- Area E: 3.2 m<br>- Area B: 3.8 m | Area E:<br>120 ×<br>40 m$^2$<br>Area B:<br>140 ×<br>60 m$^2$ |
| Bellutagi G.S. et al. [35] | BLE, QR code and inertial sensors | QR code, iBeacon | High accuracy<br>Low maintenance price<br>Medium infrastructure price | – |
| Chirakkal V. V. et al. [36] | QR code and inertial sensors | PDR | Average error:<br>0.64 m | – |
| Real Ehrlich C. e Blankenbach J. [37] | Inertial sensors and Building Information Modeling (BIM) | Sequential Monte Carlo (SMC), WLAN fingerprinting, RSS BLE, Magnetic Anomaly (MA) | Average error:<br>(Sony Z5/Google Pixel 2 XL)<br>11.19 m/11.78 m +<br>WLAN fingerprint:<br>7.22 m/7.03 m + BLE beacon:<br>1.98 m/3.27 m +<br>MA:18.2 m/10.45 m<br>+ WLAN FP + BLE:<br>2.95 m/3.25 m + MA<br>+ BLE:<br>2.24 m/2.06 m<br>Together:<br>2.54 m/3.28 m | 83.325 ×<br>50.50 m$^2$ |
| Park J.W. et al. [38] | BLE, BIM and inertial sensors | RSSI BLE, PDR, BIM | Average error/(Standard deviation)<br>1$^{st}$ SCENARIO:<br>1.15 m/(0.72 m)<br>2$^{nd}$ SCENARIO:<br>2.03 m/(1.22 m) | 27.4 ×<br>39 m$^2$ |

**Table 5.** Computer-vision systems applications for indoor navigation.

| Authors | Method | System | Performance | Test size |
|---|---|---|---|---|
| Elloumi W. et al. [39] | Computer-vision, inertial sensors | Harris-Based Matching, ZUPT (Zero Velocity Update) | Average error: Computer-vision: from 0.519 m to 1.503 m Sensors: from 1.276 m to 4.146 m | Different route lengths |
| Zhou Y. et al. [40] | Computer-vision, BIM | Visual matching between artificial target (BIM) and smartphones' cameras | 0.01 m | Few meters |
| Huang G. et al. [41] | WiFi, visual sensors | Wi-Fi fingerprint | Average error: <0.5 m | Area 12 000 m$^2$ |
| Kunhoth J. et al. [42] | Computer-vision with BLE, trained deep learning computer-vision (CamNav) and QR code computer-vision (QRNav) comparison | Scene analysis with smartphones' camera; BLE fingerprinting and multilateration; deep learning with Tensorflow; QR code | Standard deviation: Route 1/Route 2 CamNav 3.1 m(0.56 m)/6.1 m(1.10 m) QRNav 3.3 m (0.48 m)/5.5m(0.84 m) BLE APP 4.3 m(0.94 m)/8.7 m(1.33 m) | – |
| Neges M. et al. [43] | Augmented Reality (AR) and inertial sensors integration | Recalibration occurs every time a natural marker is identified | Good accuracy | – |

## 3 Results

The ideal solution for the main scope of this project, a RTLS for indoor navigation of patients and general users throughout Le Scotte Hospital of Siena, would be the adoption of a hybrid system with Augmented Reality (AR) techniques with QR-code identification

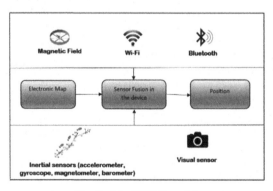

**Fig. 1.** Ideal RTLS model.

wherever a junction is, in order to re-calibrate the position, magnetic field systems along the hallways, WiFi and BLE to better perform in terms of accuracy, and PDR algorithms (Fig. 1).

Unfortunately, this solution results in a very expensive system, because it would require a massive WiFi coverage and iBeacons for iOS compatibility. Furthermore, the system itself would also need frequent re-calibration due to the normal moving of metal medical devices through the facility [44]. A better solution, in terms of price/quality ratio is the adoption of an AR system wherever the environment is wider

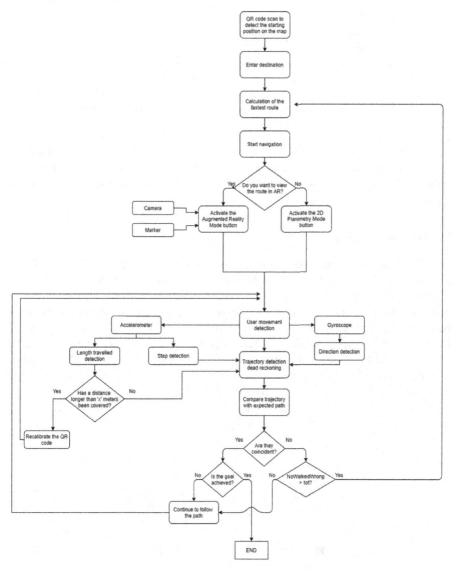

**Fig. 2.** Flowchart of a hybrid solution with AR, PDR and QR-code.

and more complex. PDR can be used along the hallways, by implementing equally spaced QR-code to perform re-calibration of devices to avoid drift-error divergency (Figs. 2 and 3).

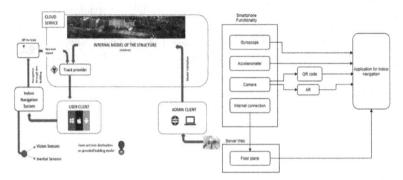

**Fig. 3.** Map-loading diagram (left) and functional diagram (right).

Two different scenarios must be taken into consideration: off-site and on-site navigation. The former helps the users to prior analyse the route of interest before reaching the hospital, while the latter guides the users to the desired destination step by step, directly in place.

### 3.1 Off-Site Navigation

For off-site navigation the best designing option is a web-site with 2D digital plan navigation and virtual touring [45]. BIM data are used to obtain 3D images of the inner structure of the hospital, while panorama pictures, which are directly attached on the 3D model, are the main inputs for a virtual tour recreation: it offers navigation by images of the site with manual scrolling. The chosen software for this particular task is Unity 3D, because it is easily programmable via Javascript and C#, a lot of online helping material is available and freely accessible, and it also has cross-platform compatibility, avoiding dedicated Android and iOS programming. In particular, Navigation Mesh (NavMesh) functionality of Unity, may be helpful to find the shortest path between two given points once a 3D model is loaded into the software (Fig. 4). BIM data can also be used to extract 2D information for digital plan navigation [46–48].

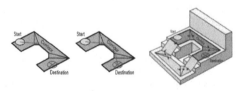

**Fig. 4.** NavMesh algorithm in Unity 3D

## 3.2  On-Site Navigation

When users are right on place, an easier way to guide them from their current position to the desired destination must be designed. During this preliminary designing phase, easy-access routing is not taken into consideration, because it requires different types of navigation according to the disabilities of the users, which is postponed to future development (Fig. 5).

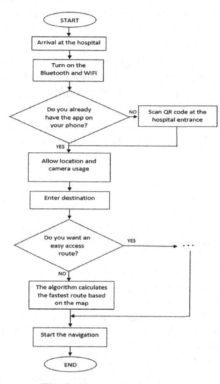

**Fig. 5.**  User action flowchart.

In this scenario, a mobile-application is the best solution, because it can easily access the hardware of the device itself (accelerometer, gyroscope, camera), which is mandatory to perform the chosen hybrid RTLS (Chapter 2). The main problem when it comes to mobile application deploying, is the different framework each Operative System (iOS and Android) is compliant to. The adoption of cross-platform deploying framework, such as Apache Cordova or Xamarin, is the best option to avoid redundant programming and maintenance. These frameworks allow natively programming in Javascript, C#, HTML5, CSS, and then to compile the code for Android and iOS by using inner libraries. Apache Cordova has been chosen among the various possibilities because it can easily communicate with the needed hardware via API, in order to perform AR, QR-code identifying and PDR algorithms.

## 4  Conclusions

This work presented different solutions for indoor Real-Time Location System. WiFi, WLAN, UWB, BLE, magnetic fields, infrareds, ultrasounds, computer-vision and PDR have all been analysed and compared in terms of accuracy, coverage, price, installation complexity and maintenance. Actually, the main scope is to design a RTLS for both on-site and off-site navigation, for hospitals and healthcare facilities. The case study the project is focuses to is Le Scotte Hospital in Siena (Italy), so that the peculiarities of the premise have also been taken into consideration when it came to choose the best-fitting solution.

The result is a hybrid system, which combined computer-vision and PDR technologies, together with simple QR-coding. A web-site with virtual touring and plain map navigation developed with Unity 3D is the solution adopted for off-site navigation, while a mobile application with actual RTLS functionalities developed by using the cross-platform framework Apache Cordova, is the chosen solution for on-site IPS.

Future works will surely consist in developing and deploying the system, testing it on the case-study hospital, and they may also include easy-access routing for people with disabilities.

## References

1. Kunhoth, J., Karkar, A., Al-Maadeed, S., et al.: Indoor positioning and wayfinding systems: a survey. Hum. Cent. Comput. Inf. Sci. **10**, 18 (2020). https://doi.org/10.1186/s13673-020-00222-0
2. Pancham, J., Millham, R., Fong, S.J.: Evaluation of real time location system technologies in the health care sector. In: Proceedings of the 2017 17th International Conference on Computational Science and Its Applications, ICCSA 2017 (2017)
3. Mainetti, L., Patrono, L., Sergi, I.: A survey on indoor positioning systems. In: 2014 22nd International Conference on Software, Telecommunications and Computer Networks, SoftCOM 2014, vol. 2014, pp. 111–120 (2014)
4. Perez-Navarro, A., Montoliu, R., Torres-Sospedra, J., Conesa, J.: Magnetic field as a characterization of wide and narrow spaces in a real challenging scenario using dynamic time warping. In: 9th International Conference on Indoor Positioning and Indoor Navigation, IPIN 2018 (2018)
5. Iadanza, E., Luschi, A., Gusinu, R., Terzaghi, F.: Designing a healthcare computer aided facility management system: a new approach. In: Badnjevic, A., Škrbić, R., Gurbeta Pokvić, L. (eds.) CMBEBIH 2019. IP, vol. 73, pp. 407–411. Springer, Cham (2020). https://doi.org/10.1007/978-3-030-17971-7_61
6. Iadanza, E., Luschi, A.: An integrated custom decision-support computer aided facility management informative system for healthcare facilities and analysis. Health Technol. **10**(1), 135–145 (2019). https://doi.org/10.1007/s12553-019-00377-6
7. Luschi, A., Marzi, L., Miniati, R., Iadanza, E.: A custom decision-support information system for structural and technological analysis in healthcare. In: Roa Romero, L. (eds.) XIII Mediterranean Conference on Medical and Biological Engineering and Computing 2013. IFMBE Proceedings, vol. 41, pp. 1350–1353. Springer, Cham (2014). https://doi.org/10.1007/978-3-319-00846-2_334

8. Ibrahim, M., Nabil, T., Halawa, H.H., ElSayed, H.M., Daoud, R.M., Amer, H.H., et al.: Fuzzy-based Wi-Fi localisation with high accuracy using fingerprinting. Int. J. Syst. Control Commun. **9**(1), 1–19 (2018)
9. Chervoniak, Y., Gorovyi, I.: Mobile indoor navigation: from research to production. In: 2019 Signal Process Symposium, SPSympo 2019, pp. 96–99 (2019)
10. Cui, Y., Zhang, Y., Wang, Z., Fu, H.: Integrated WiFi/MEMS indoor navigation based on searching space limiting and self-calibration. Arab. J. Sci. Eng. **45**(4), 3015–3024 (2019). https://doi.org/10.1007/s13369-019-04249-z
11. Li, X.J., Bharanidharan, M.: RSSI fingerprinting based iphone indoor localization system without Apple API. Wirel. Pers. Commun. **112**(1), 61–74 (2019). https://doi.org/10.1007/s11277-019-07015-4
12. Jachimczyk, B.: Real-Time Locating Systems for Indoor Applications: The Methodological Customization Approach. Blekinge Institute of Technology, Sweden (2019)
13. Chen, L., Wu, J., Yang, C.: MeshMap: a magnetic field-based indoor navigation system with crowdsourcing support. IEEE Access **8**, 39959–39970 (2020)
14. Bhattarai, B., Yadav, R.K., Gang, H.S., Pyun, J.Y.: Geomagnetic field based indoor landmark classification using deep learning. IEEE Access. **7**, 33943–33956 (2019)
15. Benedetti, M., Bononi, L., Bedogni, L.: Un algoritmo di geolocalizzazione indoor basato su magnetismo. Alma Mater Studiorum, Università di Bologna (2016)
16. Möller, A., Kranz, M., Huitl, R., Diewald, S., Roalter, L.: A mobile indoor navigation system interface adapted to vision-based localization. In: Proceedings of the 11th International Conference on Mobile and Ubiquitous Multimedia, MUM 2012 (2012)
17. Iadanza, E., Turillazzi, B., Terzaghi, F., Marzi, L., Giuntini, A., Sebastian, R.: The streamer European project. Case study: Careggi Hospital in Florence. In: Lacković, I., Vasic, D. (eds.) 6th European Conference of the International Federation for Medical and Biological Engineering. IP, vol. 45, pp. 649–652. Springer, Cham (2015). https://doi.org/10.1007/978-3-319-11128-5_162
18. Iadanza, E., Dori, F., Miniati, R., Corrado, E.: Electromagnetic interferences (EMI) from Active RFid on critical care equipment. In: Bamidis, P.D., Pallikarakis, N. (eds.) XII Mediterranean Conference on Medical and Biological Engineering and Computing 2010. IFMBE Proceedings, vol. 29. Springer, Heidelberg (2010). https://doi.org/10.1007/978-3-642-13039-7_251
19. Iadanza, E., Dori, F.: Custom active RFId solution for children tracking and identifying in a resuscitation ward. In: Proceedings of the 31st Annual International Conference of the IEEE Engineering in Medicine and Biology Society: Engineering the Future of Biomedicine, EMBC 2009, pp. 5223–5226 (2009). Art. no. 5333497
20. Gholamhosseini, L., Sadoughi, F., Safaei, A.: Hospital real-time location system (a practical approach in healthcare): a narrative review article. Iran J. Pub. Health **48**(4), 593–602 (2019)
21. Sadowski, S., Spachos, P.: Comparison of RSSI-based indoor localization for smart buildings with Internet of Things. In: 2018 IEEE 9th Annual Information Technology, Electronics and Mobile Communication Conference, IEMCON 2018, vol. 2018, pp. 24–29 (2018)
22. Wang, X., Gao, L., Mao, S.: BiLoc: bi-modal deep learning for indoor localization with commodity 5 GHz WiFi. IEEE Access **5**, 4209–4220 (2017)
23. Joseph, R., Sasi, S.: Indoor positioning using WiFi fingerprint, pp. 1–3 (2018)
24. Yu, J., Na, Z., Liu, X., Deng, Z.: WiFi/PDR-integrated indoor localization using unconstrained smartphones. EURASIP J. Wirel. Commun. Netw. **2019**(1), 1–13 (2019). https://doi.org/10.1186/s13638-019-1365-9
25. Abdulkarim, H.D., Sarhang, H.: Normalizing RSS values of Wi-Fi access points to improve an integrated indoors smartphone positioning solutions. In: Proceedings of the 5th International Engineering Conference, IEC 2019, vol. 2019, pp. 171–176 (2019)

26. Tian, Z., Fang, X., Zhou, M., Li, L.: Smartphone-based indoor integrated WiFi/MEMS positioning algorithm in a multi-floor environment. Micromachines **6**(3), 347–363 (2015)

27. Selamat, M.H., Narzullaev, A.: Wi-Fi signal strength vs. magnetic fields for indoor positioning systems. Eurasian J. Math. Comput. Appl. **2**(2), 122–133 (2014)

28. Bellutagi, G.S., Priya, D., Amith, M.: Indoor navigation using QR code, Pedometer and IBeacon. In: 2nd International Conference on Computational Systems and Information Technology for Sustainable Solutions, CSITSS 2017, pp. 36–43 (2017)

29. Chirakkal, V., Park, M., Han, D.S., Shin, J.-H.: An efficient and simple approach for indoor navigation using smart phone and QR code (2014)

30. Real Ehrlich, C., Blankenbach, J.: Indoor localization for pedestrians with real-time capability using multi-sensor smartphones. Geo-Spatial Inf. Sci. **22**(2), 73–88 (2019)

31. Park, J.W., Chen, J., Cho, Y.K.: Self-corrective knowledge-based hybrid tracking system using BIM and multimodal sensors. Adv. Eng. Inf. **1**(32), 126–138 (2017)

32. Lee, N., Ahn, S., Han, D.: AMID: accurate magnetic indoor localization using deep learning. Sensors **18**(5), 1–6 (2018)

33. Ning, F.S., Chen, Y.C.: Combining a modified particle filter method and indoor magnetic fingerprint map to assist pedestrian dead reckoning for indoor positioning and navigation. Sensors (Switz.) **20**(1), 185 (2020)

34. Li, Y., Zhuang, Y., Lan, H., Zhou, Q., Niu, X., El-Sheimy, N.: A Hybrid WiFi/magnetic matching/PDR approach for indoor navigation with smartphone sensors. IEEE Commun. Lett. **20**(1), 169–172 (2016)

35. Ashraf, I., Hur, S., Shafiq, M., Kumari, S., Park, Y.: GUIDE: smartphone sensors-based pedestrian indoor localization with heterogeneous devices. Int. J. Commun. Syst. **32**, 1–9 (2019)

36. Chen, Y., Zhou, M., Zheng, Z.: Learning sequence-based fingerprint for magnetic indoor positioning system. IEEE Access **7**, 163231–163244 (2019). https://doi.org/10.1109/ACCESS.2019.2952564

37. Shu, Y., Bo, C., Shen, G., Zhao, C., Li, L., Zhao, F.: Magicol: indoor localization using pervasive magnetic field and opportunistic WiFi sensing. IEEE J. Sel. Areas Commun. **33**(7), 1443–1457 (2015)

38. Li, P., Yang, X., Yin, Y., Gao, S., Niu, Q.: Smartphone-based indoor localization with integrated fingerprint signal. IEEE Access **8**, 33178–33187 (2020)

39. Elloumi, W., Latoui, A., Canals, R., Chetouani, A., Treuillet, S.: Indoor pedestrian localization with a smartphone: a comparison of inertial and vision-based methods. IEEE Sens. J. **16**(13), 5376–5388 (2016)

40. Zhou, Y., Li, G., Wang, L., Li, S., Zong, W.: Smartphone-based pedestrian localization algorithm using phone camera and location coded targets. In: Proceedings of 5th IEEE Conference on Ubiquitous Positioning, Indoor Navigation and Location-Based Services, UPINLBS 2018, pp. 1–7 (2018)

41. Huang, G., Hu, Z., Wu, J., Xiao, H., Zhang, F.: WiFi and vision integrated fingerprint for smartphone-based self-localization in public indoor scenes. IEEE Internet Things J. **4662**(c), 1–16 (2020)

42. Kunhoth, J., Karkar, A., Al-Maadeed, S., Al-Attiyah, A.: Comparative analysis of computer-vision and BLE technology based indoor navigation systems for people with visual impairments. Int. J. Health Geogr. **18**(1), 1–18 (2019)

43. Neges, M., Koch, C., König, M., Abramovici, M.: Combining visual natural markers and IMU for improved AR based indoor navigation. Adv. Eng. Inf. **31**, 18–31 (2017)

44. Iadanza, E., Marzi, L., Dori, F., Biffi Gentili, G., Torricelli, M.C.: Hospital health care offer. A monitoring multidisciplinar approach. In: Magjarevic, R., Nagel, J.H. (eds.) World Congress on Medical Physics and Biomedical Engineering 2006. IFMBE Proceedings, vol. 14, pp. 3685–3688. Springer, Heidelberg (2007). https://doi.org/10.1007/978-3-540-36841-0_933
45. Snopková, D., Švedová, H., Kubíček, P., Stachoň, Z.: Navigation in indoor environments: does the type of visual learning stimulus matter? ISPRS Int. J. Geo-Inf. **8**(251), 26 (2019)
46. Iadanza, E., Luschi, A., Ancora, A.: Bed management in hospital systems. In: Lhotska, L., Sukupova, L., Lacković, I., Ibbott, G.S. (eds.) World Congress on Medical Physics and Biomedical Engineering 2018. IP, vol. 68/3, pp. 313–316. Springer, Singapore (2019). https://doi.org/10.1007/978-981-10-9023-3_55
47. Luschi, A., Monti, M., Iadanza, E.: Assisted reproductive technology center design with quality function deployment approach. In: Jaffray, D.A. (ed.) World Congress on Medical Physics and Biomedical Engineering, June 7-12, 2015, Toronto, Canada. IP, vol. 51, pp. 1587–1590. Springer, Cham (2015). https://doi.org/10.1007/978-3-319-19387-8_386
48. Iadanza, E., Luschi, A., Merli, T., Terzaghi, F.: Navigation algorithm for the evacuation of hospitalized patients. In: Lhotska, L., Sukupova, L., Lacković, I., Ibbott, G.S. (eds.) World Congress on Medical Physics and Biomedical Engineering 2018. IP, vol. 68/3, pp. 317–320. Springer, Singapore (2019). https://doi.org/10.1007/978-981-10-9023-3_56

# Sequential Pattern Mining Method for Predictive Maintenance of Large Mining Trucks

Abdulgani Kahraman[1]($\boxtimes$), Mehmed Kantardzic[1] ⓘ, M. Mustafa Kahraman[2] ⓘ, and Muhammed Kotan[3] ⓘ

[1] Department of Computer Engineering and Computer Science, University of Louisville, 132 Eastern Pkwy, Louisville, KY 40292, USA
a0kahr01@louisville.edu

[2] Department of Mining Engineering, Gumushane University, 29100 Gumushane, Turkey

[3] Department of Information Systems Engineering, Sakarya University, Esentepe Campus, 54050 Serdivan, Sakarya, Turkey

**Abstract.** In recent decades, various solutions had been sought for reducing operating costs while increasing the production of minerals in mining operations. Equipment health monitoring technologies had been used for monitoring and increasing the availability of machines. However, the data obtained from these technologies had only been used for monitoring the equipment health, and not for the prediction of failures. In this paper, it was relied on alarms and signals collected through real-time health monitoring technologies for predicting crucial mining truck failures. Sequential Pattern Mining (SPM) Method for Predictive Maintenance had been developed and implemented as a methodology to discover which group of alarms and signals might be related to specific truck failures. The results indicate that the SPM method is able to detect machine failures of trucks with high accuracy with an average 96%. The proposed methodology may reduce the maintenance time, and the expenditures caused by truck breakdowns in the mining industry.

**Keywords:** Sequential pattern mining · Predictive maintenance · Mining trucks · Mine equipment

## 1 Introduction

Since recent decades, remote access technology has allowed the maintenance teams to monitor and control the equipment from distant locations via internet, phone, satellite links, or radio. The use of the internet for failure detection and troubleshooting of mining equipment is becoming a standard in the industry [1]. Spokane Mining Research Division had partnered with a concrete company for developing and implementing wireless Internet of Things (IoT) solutions in order to monitor the equipment and processes on a real-time basis [2]. The machine failures identified by a few pointers, and the requirements led to high expenses in the maintenance systems that were continuously being evaluated for cost reduction, and for keeping the machines in excellent working condition [3, 4]. Maintenance activities on relevant machines and equipment are regularly

© Springer Nature Switzerland AG 2021
J. Hasic Telalovic and M. Kantardzic (Eds.): MeFDATA 2020, CCIS 1343, pp. 126–136, 2021.
https://doi.org/10.1007/978-3-030-72805-2_9

performed initially by the integration of maintenance and process engineering functions, and then by application of the machine and hardware, and finally by proactive actions [5] including preventive and predictive maintenance.

Failure detection and maintenance operations of machinery are large scaled, primarily due to the variety of production systems. Each year, many papers, covering theories and applicable methods, are being published on this subject in academic journals [3–8], and the issue is being addressed at conferences, and in technical reports [9].

Summit and Halomoan had calculated the life expectancy of suspensions of trucks by implementing Pareto analysis. According to their research, most of the suspensions can be replaced earlier than their expected life due to bad road conditions, and poor haulage [10]. Another paper on trucks by Marinelli et al. from Greece had made use of ANN-based predictive model, and it had used the capacity, age, mileage, and maintenance level data obtained from 126 earthmoving trucks. Their model had reached an accuracy of above 92% for predicting the conditions of trucks [11]. Similarly, the Random Forest method had been used by Prytz et al. for the efficient maintenance of European Volvo trucks based on maintenance records. The results had revealed a link between trucks' mileage and age, and engine operation hours, and repair of air compressors [12].

Dindarloo and Siami-Irdemoosa had used k-means clustering and support vector machine techniques for predicting the failures of mining shovels with an accuracy above 75% [13]. Peng and Vayenas had presented a study on the implementation of genetic algorithm on Underground Mining Equipment as a case study for predictive maintenance. They had assumed that the failures of the mining equipment were being caused by an array of factors, and they had followed the biological evolution theory; however, significant impacts of chronological sequence couldn't be found on the results [14].

Sequential Pattern Mining (SPM) is one of the most important mining techniques for analyzing big data [15]. SPM has a broad field of application such as natural disasters, sales record analysis, marketing strategies, shopping sequences, medical treatments, and DNA sequences etc. [16].

The purpose of the present research is to detect patterns, that may be indicators of truck-specific breakdowns, based on past readings of equipment health sensors.

The paper consists of the following sections: Sect. 1 introduces the topic and gives the background in the field of mining. And Sect. 2 consists of the details on the dataset, and on proposed sequential pattern mining method for predictive maintenance. The results of the real-world example are shown in Sect. 3. The conclusion, and information on future studies are given in Sect. 4.

## 2  Experiments

### 2.1  Dataset Details

The dataset was collected for about ten months from 11 mining trucks which were working for 24 h a day and seven days a week in North America. The elements of dataset, consisting of trucks' status, signals, changes, and alarms, were recorded by chips and sensors.

The dataset summarized in Fig. 1 indicates trucks' downtime percentages over quartiles, and the trucks' identification numbers. The dataset includes partial data for eleven trucks for the period between 2010 and 2011.

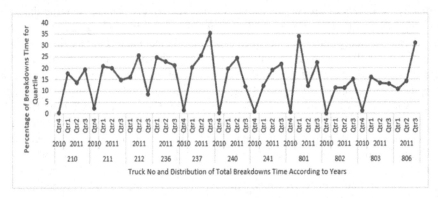

**Fig. 1.** Total breakdown times as a percentage according to date and truck numbers

The downtimes have substantial impact on trucks' operating times. Several machines had been inactive for more than 30% of the time in various quartiles as shown in Fig. 1. Thus, detecting unusual patterns prior to failure of machine is crucial for ensuring efficiency.

## 2.2  Methodology.

SPM focuses on finding patterns that occur consecutively in a database, and these patterns are related to time or other variables. The aim of pattern mining is to discover useful, recent, and unforeseen patterns in databases.

In the present paper, we used the modified Generalized Sequential Pattern (GSP) algorithm, and we called our specific method SPMPM (Sequential Pattern Mining for Predictive Maintenance). The GSP algorithm [17] generates all possible k-sequences from frequent k−1 sequences by joining any two sequences in a process which is called candidate generation, and then tests the number of candidates that reach the threshold value. Each sequential pattern between the same failure codes, and the sequential patterns for the recent five shifts prior to the occurrence of specific machine failure were revealed. Furthermore, these groups of sequential patterns were found to have maximum size of patterns as 2, 3, 4, and 5.

Our motivation was that most of readings of machine sensors were following a pattern of an incidence arising close to their breakdowns. Based on this approach, various combinations of shifts prior to occurrence of specific failures were examined. As a result, the recent five shifts were found as the optimal shift size. It may be required to adjust the maximum size of the groups, and the shift size according to data.

As a pseudocode for SPMPM algorithm:

---

$E \leftarrow \{e_i \mid i \in Z^+, i \leq N_e, i: event\ number, N_e: total\ number\ of\ different\ events\}$

$F \leftarrow \{f_j \mid j \in Z^+, j \leq N_f, j: fault\ number, N_f: total\ number\ of\ different\ faults\}$

$D \leftarrow \left\{ \begin{array}{l} \{d_1, \ldots, d_n\} \mid d \in \{E \cup F\}, \quad is\ the\ collection\ of\ events\ and \\ \hspace{3cm} faults\ sorted\ by\ their\ occurance\ time \end{array} \right\}$

$S_j$ : set of event sequences between all occurrences of fault type

$SS_j$ : $S_j$ with specified shift size before faults

$SP$: whole survival patterns for $S_j$ and $SS_j$

$j$: fault number, $N_f$ : total number of different faults

$f_j^n$: $(n)^{th}$ occurance of fault $j$, $\sigma$: shift size , shift: 12 hours

 for $j=1$ to $N_f$

   for $n=1$ to max. occurrence number of fault $f_j$

$$S_j \leftarrow \left\{ \begin{array}{c} \{U\ \forall e \mid e \in D, \\ occurance\ time\ of\ e\ is\ between\ (f_j^{n-1}, f_j^n) \end{array} \right\}$$

$$SS_j \leftarrow \left\{ \begin{array}{c} \{U\ \forall e \mid e \in D\ and\ SS_j \subset S_j, \\ occurance\ time\ of\ e\ is\ between\ \left((f_j^n - shift * \sigma), f_j^n\right) \end{array} \right\}$$

   end

   $C_k$: set of the candidates for k-sequences

   $P_k$: set of the frequent k-sequence

   for each $S_j$ & $SS_j$

     $C_1 \leftarrow$ init-pass (sequences in $S_j$ or $SS_j$)

     $P_1 \leftarrow \{ a : a \in C_1$ and frequency exceeds the threshold $\}$

     $k=2$

     do (while $P_{k-1}$ is not Null and $k<=5$)

       $C_k \leftarrow$ Generate $C_k$ candidate sets from $P_{k-1}$

       do (For all $s$ input sequences in $S_j$ or $SS_j$)

        if $s$ supports $a$, increase the count of every $a$ in $C_k$

       end

       $P_k \leftarrow \{ a : a \in C_k$ and frequency exceeds the threshold$\}$

       $k=k+1$

     end

     $SP(S_j$ or $SS_j) \leftarrow \{U\ P_k$ :union of All $P_k$ survival patterns, $k \in Z^+, k \in [1,5]\}$

   end

   for each common survival 'a' pattern in $SP(S_j)$ and $SP(SS_j)$

   $PatternConfidence = \left\{ \dfrac{frequency\ count\ of\ a\ in\ SP(SS_j)}{frequency\ count\ of\ a\ in\ SP(S_j)} \mid a \in SP(SS_j) \cap SP(S_j) \right\}$

   end

  end

---

Initially, the dataset was purged from missing data as a pre-processing step, and then sequential patterns between two consecutive identical failures were discovered. Following the discovery of patterns, the model tried to find a relationship between specific patterns, and certain machine breakdown codes. For this purpose, sequential patterns in the recent five shifts were discovered prior to occurrence of specific machine failures. Lastly, occurrence times of these patterns for the last five shifts, and all occurrence times between identical consecutive failures were divided, and thus we obtained the confidence values for each pattern group.

## 3   Experimental Results

In the dataset, each shift refers to 12 h. SPMPM was implemented for the recent five shifts. Sequential pattern groups with 100% confidence value were selected. Furthermore, the detection rates were calculated based on Eq. (1) for certain machine failures. The equation for the detection rate is given below:

$$Detection\ Rate = \frac{Number\ of\ Detected\ Sequential\ Groups}{Total\ Number\ of\ Occurrence\ of\ Specific\ Failure} \tag{1}$$

According to Eq. 1, the detection rate is calculated by dividing the number of detected sequential groups to the total number of specific failures for the consecutive identical machine breakdowns.

Mechanical failures are subject to higher costs, and they are related to alarms and signals at a higher extent. The mechanical failure codes were chosen as 1104, 1140, and 1143 for this paper.

### 3.1   Pattern Results for Failure Code 1104

Table 1 indicates the recent five shifts' sequential patterns with a 100% confidence value for failure code 1104. In this table, the first column indicates the pattern groups, and the second column indicates group numbers, and the last three columns indicate pattern sizes, and counts as per all the shifts and as per the recent 5 shifts.

**Table 1.** Distribution of sequence pattern groups with a confidence value 100% as per order of breakdowns for failure code 1104 in the recent five shifts

| Sequential pattern groups | Sequential patterns groups number | Pattern size | Number of patterns in all shifts | Number of patterns in last 5 shift |
|---|---|---|---|---|
| 0E010005,0E010004 | s1 | 2 | 2 | 2 |
| 3E010000,0E01FFE1,0E01000B,0E01000A,0E010000 | s2 | 5 | 2 | 2 |

(continued)

**Table 1.** (*continued*)

| Sequential pattern groups | Sequential patterns groups number | Pattern size | Number of patterns in all shifts | Number of patterns in last 5 shift |
|---|---|---|---|---|
| 3E010002,0E01000B,0E01000A,0E010009 | s3 | 4 | 2 | 2 |
| 0E010001,3E010001 | s4 | 2 | 2 | 2 |
| 0E010003,0E010009,0E010008,0E010007 | s5 | 4 | 4 | 4 |
| 0E010008,0E01000B,0E01000A,0E01FFE0 | s6 | 4 | 2 | 2 |
| 0E01000A,0E010009,0E010008 | s7 | 3 | 2 | 2 |
| 0E01FFE0,0E01000A | s8 | 2 | 2 | 2 |
| 0E01FFE0,0E01FFE1,0E010009,0E010008,3E010002 | s9 | 5 | 2 | 2 |
| 0E01FFE0,3E010002,0E01FFE1 | s10 | 3 | 2 | 2 |
| 0E01FFE1,0E010008,0E010009,0E010000 | s11 | 4 | 2 | 2 |
| 0E01FFE1,0E01000A | s12 | 2 | 2 | 2 |
| 3E010002,0E010002 | s13 | 2 | 2 | 2 |

In Fig. 2a and Fig. 2b, detected sequence groups were distributed by occurrence amount of failure code 1104. Furthermore, they show which pattern groups were detected for each occurrence.

The detection rate is calculated by dividing number of detected sequences (illustrated as a yes for each fault no in Fig. (2b)) to the total number of failures. In this example, the number of detected sequences (7) was divided to the total number of failures (7), and that had resulted in a 100% detection rate.

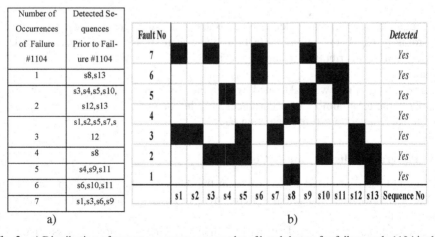

a)                                        b)

**Fig. 2. a)** Distribution of sequence groups as per order of breakdowns for failure code 1104 in the recent five shifts. **b)** Graphical representation of the distribution of sequence groups as per order of breakdowns for failure code 1104 in the recent five shifts.

### 3.2  Pattern Results for Failure Code 1140

For failure 1140, there were 12 different sequential pattern groups, and failure 1140 had occurred for six times in the dataset. The maximum pattern size was five, and the confidence value was 100%.

**Table 2.** Distribution of sequence pattern groups with a confidence value 100% as per order of breakdowns for failure code 1140 in the recent five shifts

| Sequential pattern groups | Sequential patterns groups number | Pattern size | Number of patterns in all shifts | Number of Patterns in last 5 shift |
|---|---|---|---|---|
| 0201FFE0,0B010009 | s1 | 2 | 2 | 2 |
| 0B010011,160169DD,16016A41,160169DC | s2 | 4 | 3 | 3 |
| 2,010,000,0B01000C | s3 | 2 | 2 | 2 |
| 16,016,465,160178B5,160178B4,16,016,464,2,010,006 | s4 | 5 | 2 | 2 |
| 0201000A,0B01000C,2,010,002,2,010,004,0B010014 | s5 | 5 | 4 | 4 |
| 0201FFE1,0201FFE0 | s6 | 2 | 4 | 4 |
| 070108CC,160178B5 | s7 | 2 | 2 | 2 |
| 0B011519,0B010000 | s8 | 2 | 2 | 2 |
| 0B011519,0B01000C,0B01000A | s9 | 3 | 2 | 2 |
| 0B01000A,160165F4 | s10 | 2 | 2 | 2 |
| 2,010,006,16,016,465,2,010,000,160178B5,2,010,005 | s11 | 5 | 2 | 2 |
| 160178B5,160178B4,2,010,005,2,010,006 | s12 | 4 | 2 | 2 |

Figure 3a and Fig. 3b are derived from Table 2, and the detection rate is calculated according to Eq. 1. For failure code 1140, there were 12 different sequential pattern groups, and its distribution is shown in Fig. 3a based on number of occurrences of failure.

The detection rate was calculated by dividing the number of detected sequences in Fig. 3a to the total number of occurrences of failure, and it was 6 for failure no 1140. And it had resulted in a 100% detection rate.

### 3.3  Pattern Results for Failure Code 1143

Failure 1143 had occurred for 17 times in the dataset. Following the application of SPMPM, 12 different sequential patterns were detected for the recent five shifts as shown in Table 3.

Figure. 4a and Fig. 4b are derived from Table 3, and the detection rate was calculated according to Eq. 1. For failure code 1143, there were 12 different sequential pattern groups, and its distribution is shown in Fig. 4a based on number of occurrences of failure.

| Occurrence No for Faults #1140 | Observed Sequences Before Faults#1140 |
|---|---|
| 1 | s2,s5,s7,s10 |
| 2 | s7 |
| 3 | s1,s3,s4,s6,s9, s11, s12 |
| 4 | s3,s6,s8,s9 |
| 5 | s2,s5 |
| 6 | s5,s10 |

a)

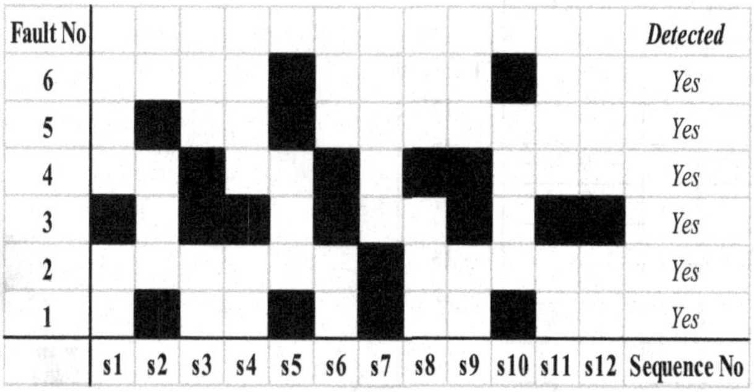

| Fault No | Detected |
|---|---|
| 6 | *Yes* |
| 5 | *Yes* |
| 4 | *Yes* |
| 3 | *Yes* |
| 2 | *Yes* |
| 1 | *Yes* |

s1 s2 s3 s4 s5 s6 s7 s8 s9 s10 s11 s12 Sequence No

b)

**Fig. 3. a)** Distribution of sequence groups as per order of breakdowns for failure code 1140 in the recent five shifts. **b)** Graphical representation of the distribution of sequence groups as per order of breakdowns for failure code 1140 in the recent five shifts.

**Table 3.** Distribution of sequence pattern groups with a confidence value 100% as per order of breakdowns for failure code 1143 in the recent five shifts

| Sequential pattern groups | Sequential patterns groups number | Pattern size | Number of patterns in all shifts | Number of patterns in recent 5 shift |
|---|---|---|---|---|
| 05010ADD,05010ADC | s1 | 2 | 2 | 2 |
| 050114B4,050114B9,050114B8,5,010,905 | s2 | 4 | 2 | 2 |
| 050116CA,050114B5 | s3 | 2 | 2 | 2 |
| 050119A0,050116CB,050116CA | s4 | 3 | 2 | 2 |
| 0501199B,0501199A,3E010002 | s5 | 3 | 2 | 2 |
| 050114B9,050114B7,050114B6 | s6 | 3 | 2 | 2 |
| 0501199D,0501199C,5,010,249,5,010,248 | s7 | 4 | 3 | 3 |
| 050119CE,050119CB | s8 | 2 | 2 | 2 |
| 3E010001,050119CB,050119CA,0501177F | s9 | 4 | 2 | 2 |
| 3E010002,3E010001,3E010002 | s10 | 3 | 4 | 4 |
| 3E010002,3E010001,3E010000,0501177F | s11 | 4 | 2 | 2 |
| 5,010,248,050114B9,050114B8 | s12 | 3 | 2 | 2 |

The detection rate was determined by dividing the number of filled rows in the Fig. 4a, which was 15 (the number of yes in Fig. 4b), to the total number of faults, which was 17, and it had resulted in about 88% detection rate for failure 1143.

**a)**

| Number of Occurrences of Failure #1143 | Detected Sequences Prior to Failure #1143 |
|:---:|:---:|
| 1 | s7 |
| 2 | s3,s6,s7,s12 |
| 3 | s7,s12 |
| 4 | s8 |
| 5 | s1,s5 |
| 6 | s4,s5 |
| 7 | - |
| 8 | s2,s3 |
| 9 | s1 |
| 10 | s8 |
| 11 | s11 |
| 12 | s9,s10 |
| 13 | s9 |
| 14 | s10 |
| 15 | - |
| 16 | s10 |
| 17 | s10 |

**b)**

| Fault No | s1 | s2 | s3 | s4 | s5 | s6 | s7 | s8 | s9 | s10 | s11 | s12 | Detected |
|:---:|:---:|:---:|:---:|:---:|:---:|:---:|:---:|:---:|:---:|:---:|:---:|:---:|:---:|
| 17 | | | | | | | | | ■ | | | | *Yes* |
| 16 | | | | | | | | | ■ | | | | *Yes* |
| 15 | | | | | | | | | | | | | *No* |
| 14 | | | | | | | ■ | | | | | | *Yes* |
| 13 | | | | | | | ■ | | | | | | *Yes* |
| 12 | | | | | | | ■ | | | | | | *Yes* |
| 11 | | | | | | | | | | ■ | | | *Yes* |
| 10 | | | | | | | ■ | | | | | | *Yes* |
| 9 | ■ | | | | | | | | | | | | *Yes* |
| 8 | | ■ | | | | | | | | | | | *Yes* |
| 7 | | | | | | | | | | | | | *No* |
| 6 | | | | | ■ | | | | | | | | *Yes* |
| 5 | ■ | | | | | | | | | | | | *Yes* |
| 4 | | | | | | | ■ | | | | | | *Yes* |
| 3 | | | | | | ■ | | | | | ■ | | *Yes* |
| 2 | | | ■ | | | ■ | | | | | ■ | | *Yes* |
| 1 | | | | | | | ■ | | | | | | *Yes* |

**Fig. 4. a)** Distribution of sequence groups as per order of breakdowns for failure code 1143 in the recent five shifts. **b)** Graphical representation of the distribution of sequence groups as per order of breakdowns for failure code 1143 in the recent five shifts.

The detection rate was 100% for failures 1104 and 1140, and 88% for failure 1143. The performance of SPMPM for all the three failures had 96% detection rate on average.

## 4   Conclusion and Future Work

In this research, three kinds of breakdown codes, which had caused higher maintenance costs and mechanical failures, were selected. First, several patterns were discovered between the two same failure codes, and they were counted. Afterwards, various patterns were discovered in the recent five shifts prior to occurrence of breakdowns. Lastly, confidence values, for the recent five shifts, were calculated, and indicated in tables and graphs. The results indicated that the detection rate had a high accuracy for the recent five shifts, and that implies that several pattern groups can be identified prior to occurrence of machine breakdowns.

The results indicate that if the sequential pattern algorithm is implemented prior to machine failures, it is possible to discover several patterns which may be related to specific breakdowns. However, this study is limited with the size of the dataset. The performance of the method is subject to change for greater datasets.

The main contribution of this paper is that it shows the usefulness of sequential pattern mining method to detect machine failures before they occur. There are several repetitive patterns in machines signals. However, if useful unusual patterns were discovered, then they can be used as a potential indicator to recognize machine breakdowns.

Furthermore, implementing other up-to-date machine learning models may provide better results for predicting machine failures. Regarding future work, a more extensive dataset will increase the accuracy of the results in discovering the patterns for the failure of mining trucks, and potentially for other machine breakdowns in other industries.

# References

1. Koellner, W.G., Brown, G.M., Rodriguez, J., Pontt, J., Cortes, P., Miranda, H.: Recent advances in mining haul trucks. IEEE Trans. Ind. Electron. **51**(2), 321–329 (2004). https://doi.org/10.1109/TIE.2004.825263
2. McNinch, M., Parks, D., Jacksha, R., Miller, A.: Leveraging IIoT to improve machine safety in the mining industry. Min. Metall. Exp. **36**(4), 675–681 (2019). https://doi.org/10.1007/s42461-019-0067-5
3. Bastos, P., Lopes, I., Pires, L.: Application of data mining in a maintenance system for failure prediction. In: Steenbergen, R., van Gelder, P., Miraglia, S., Vrouwenvelder, A. (eds.) Safety, Reliability and Risk Analysis: Beyond the Horizon, pp. 933–940. CRC Press (2013). https://doi.org/10.1201/b15938-138
4. Abbasi, T., Lim, K.H., Rosli, N.S., Ismail, I., Ibrahim, R.: Development of predictive maintenance interface using multiple linear regression. In: International Conference on Intelligent and Advanced System, ICIAS 2018, pp. 1–5 (2018). https://doi.org/10.1109/ICIAS.2018.8540602
5. Bastos, P., Lopes, I. da S., Pires, L.: A maintenance prediction system using data mining techniques. In: World Congress on Engineering 2012, vol. III, pp. 1448–1453 (2012)
6. Rezig, S., Achour, Z., Rezg, N.: Using data mining methods for predicting sequential maintenance activities. Appl. Sci. **8**, 2184 (2018). https://doi.org/10.3390/app8112184
7. Alimian, M., Saidi-Mehrabad, M., Jabbarzadeh, A.: A robust integrated production and preventive maintenance planning model for multi-state systems with uncertain demand and common cause failures. J. Manuf. Syst. **50**, 263–277 (2019). https://doi.org/10.1016/j.jmsy.2018.12.001
8. Spiegel, S., Mueller, F., Weismann, D., Bird, J.: Cost-sensitive learning for predictive maintenance, pp. 1–18 (2018)
9. Murakami, T.: Development of vehicle health monitoring system (VHMS/WebCARE) for large-sized construction machine. Construction **48**, 15–21 (2002)
10. Summit, R.A., Halomoan, D.: Reliability modelling for maintenance scheduling of mobile mining equipment. ANZIAM J. **55**, 526 (2015). https://doi.org/10.21914/anziamj.v55i0.7863
11. Marinelli, M., Lambropoulos, S., Petroutsatou, K.: Earthmoving trucks condition level prediction using neural networks. J. Qual. Maint. Eng. **20**, 182–192 (2014). https://doi.org/10.1108/JQME-09-2012-0031
12. Prytz, R., Nowaczyk, S., Rögnvaldsson, T., Byttner, S.: Predicting the need for vehicle compressor repairs using maintenance records and logged vehicle data. Eng. Appl. Artif. Intell. **41**, 139–150 (2015). https://doi.org/10.1016/j.engappai.2015.02.009
13. Dindarloo, S.R., Siami-Irdemoosa, E.: Data mining in mining engineering: results of classification and clustering of shovels failures data. Int. J. Mining Reclam. Environ. **31**(2), 105–118 (2017). https://doi.org/10.1080/17480930.2015.1123599

14. Peng, S., Vayenas, N.: Maintainability analysis of underground mining equipment using genetic algorithms: case studies with an LHD vehicle. J. Min. **2014**, 1–10 (2014). https://doi.org/10.1155/2014/528414

15. Gan, W., Lin, J.C.-W., Fournier-Viger, P., Chao, H.-C., Yu, P.S.: A survey of parallel sequential pattern mining. ACM Trans. Knowl. Discov. Data **13**, 34 (2018)

16. Ubaidulla, D., Sushmitha, B.S., Vanitha, T.: A study on mining sequential pattern in time series data. Int. J. Latest Trends Eng. Technol., 374–378 (2017). Special Issue SACAIM. E-ISSN 2278-621X

17. Srikant, R., Agrawal, R.: Mining sequential patterns: generalizations and performance improvements. In: Apers, P., Bouzeghoub, M., Gardarin, G. (eds.) Advances in Database Technology, EDBT 1996, pp. 1–17. Springer Berlin Heidelberg, Berlin, Heidelberg (1996). https://doi.org/10.1007/BFb0014140

# Natural Language Processing

# Classification and Analysis of Personal and Commercial CBD Tweets

Jason S. Turner[1], Mehmed M. Kantardzic[1]($\boxtimes$), and Rachel Vickers-Smith[2]($\boxtimes$)

[1] University of Louisville, Louisville, KY 40292, USA
{jsturn04,mehmed.kantardzic}@louisville.edu
[2] University of Kentucky, Lexington, KY 40506, USA
rachel.vickers@uky.edu

**Abstract.** This study analyzes the differences in terms regarding cannabidiol (CBD) expressed by commercial sellers and personal users on Twitter. It demonstrates that data from social networks can be used by public health and medical researchers to compare the medical conditions targeted by those selling loosely-regulated substances such as CBD against the medical conditions that patients themselves are commonly treating with CBD. We collected 567,850 tweets by searching Twitter with the Tweepy Python package using the terms CBD and cannabidiol, and annotated a sample of 5,496 tweets to distinguish between personal use CBD tweets and commercial/sales-related CBD tweets. We used this sample to train two binary text classifiers to create two corpora of 169,876 personal use and 148,866 commercial/sales. Using medical, standard, and slang dictionaries, we then identified and compared the most frequently occurring medical conditions, symptoms, side effects, body parts, and other substances referenced in both corpora.

**Keywords:** Text mining · Text classification · Cannabis

## 1 Introduction

Cannabis is often presented as a medication in public policy debates. CBD in particular has grown into a multi-billion dollar industry and is being used to treat various conditions, including epilepsy and other neurological disorders, insomnia, and some mental illnesses. Ingesting CBD does not have any psychoactive side-effects, unlike the other well known chemical found in cannabis, tetrahydrocannabinol (THC). CBD remains unregulated by the FDA and has not been subjected to the same trials as most medications. In fact, the FDA has only approved one cannabis-derived medication and three cannabis-related drugs to date - all of which require a prescription - and it has not given approval to market cannabis as a safe and effective drug for treating any disease.

Since CBD-based medications and nutritional supplements have not been subjected to the same trials for efficacy and safety as other medications, there

© Springer Nature Switzerland AG 2021
J. Hasic Telalovic and M. Kantardzic (Eds.): MeFDATA 2020, CCIS 1343, pp. 139–150, 2021.
https://doi.org/10.1007/978-3-030-72805-2_10

could be imbalances between the medical conditions CBD is marketed to cure or alleviate and the rationale for personal use of CBD [15]. Twitter is a useful platform to track this potential imbalance as it provides a large corpus of both personal and commercial tweets regarding CBD.

Thus, we propose a framework for the use of text mining in social networks that can help public health experts understand the differences between personal and commercial claims about unregulated substances. There is a practical advantage to this framework: the data is readily available, easy to access, and inexpensive to use in comparison to running surveys and/or utilizing data from governments and health providers.

To demonstrate the usefulness of this framework, in this study we distinguished between tweets reflecting personal CBD use and tweets reflecting the sales, promotion, and or commercialization of CBD. The two resulting corpora of tweets were analyzed for contextual terms such a rationale for use including specific regions of the body, symptoms/medical conditions mentioned, side effects experienced, and other substances mentioned in the CBD tweets. We were able to use text classification to identify two specific types of CBD tweets. Our approach also allowed us to identify medically-related terms that are being used in online CBD marketing, in relation to the medically-related terms referenced by individuals taking CBD.

## 2   Related Work

The cannabis plant has been used as a medication for centuries but the use of cannabis was criminalized in the United States in 1937. However, beginning in the 1990's states began allowing the medical use of cannabis even though the plant remained illegal at the federal level [12]. As more states relaxed their cannabis policies, public interest in cannabis evolved to embrace cannabidiol (CBD). CBD is an active chemical found in variants of the cannabis plant that does not have psychoactive side-effects, unlike the tetrahydrocannabinol (THC) variants of the cannabis plant [2].

Despite their varying legal statuses, researchers have been able to study the potential medical benefits of cannabis and CBD. Palmieri et al. observed promising results in CBD as a treatment for inflamed skinned conditions and scars [9]. There have also been many studies conducted on CBD as a treatment for anxiety and sleep disorders [6,7,10,11]. Multiple studies have been conducted on CBD as a pain reliever as well [1,3,14]. CBD has also been studied for both the treatment of cancer and cancer side-effects [5,8,14]. Despite this preliminary clinical evidence, these studies have not been large-scale and CBD is not a widely accepted or approved treatment by the United States FDA.

There have also been studies using the internet and social media to gather information about the personal and commercial discourses around CBD. Narayanan et al. made use of Internet-based data sources to examine CBD trends by examining Google searches where it was observed that interest in CBD oil increased significantly between the years 2014 to 2018 [8]. Tran and Kavuluru used cannabidiol (CBD)-related posts from Reddit and comments submitted to the FDA regarding these posts to examine the conditions that are commonly being treated by CBD [13]. The researchers in this study examined both corpora of texts for medical conditions and methods of use in posts and comments using the term "CBD," along with any indication of therapy implied in the two corpora.

Our proposed framework extends the existing CBD research by further examining the perceptions of CBD through online discussions by comparing the terms of tweets that reflect personal use of CBD and tweets that reflect sales and/or promotion of CBD. This approach provides the ability to examine which terms are being used either proportionally or disproportionally. Our methods can be applied to other research that involves analyzing trends in the consumption and advertising of unregulated substances. Figure 1 shows the workflow of tweet collection and classification.

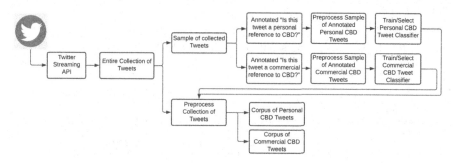

**Fig. 1.** Workflow of tweet collection and classification

## 3   Methodology

We collected tweets from the Twitter public stream using the Tweepy Python package. The Twitter stream provides access to approximately 1% of the public tweets as they are created. Our data collection ran from October 7, 2019 to January 26, 2020 using the search terms CBD and cannabidiol. We also set filters to collect only original tweets (i.e. no retweets) and that were written in English. For each tweet that we collected, we kept the full-length tweet text, the time in which the tweet was created, and any geographic information when provided. When available, we also retained the following user information: unique user identifier, screen name, name, user-provided location from their profile, user time zone, when the Twitter account was created, and the number of friends of followers at that time, as well as user-provided description from their profile. This analysis primarily focuses on the full-length tweet text. The resulting collection consisted of a dataset of 567,850 tweets.

In order to separate the personal and commercial-related CBD tweets from our collection of 567,850 tweets, we built two binary classifiers trained on a sample of 5,496 tweets. This sample was based on our preliminary experimentation of the data. These classifiers were used to distinguish between personal and commercial CBD-related tweets. To train these classifiers, each tweet in the sample we manually labeled the tweet as either a personal CBD-related tweet or a non-personal CBD-related tweet, and either a commercial CBD-related tweet or a non-commercial CBD-related tweet, according to the content in their full text, which consisted of a maximum of 280 characters. Table 1 provides some examples of the personal and commercial CBD-related tweets, as well as examples of the erroneous tweets but which are in neither category that also reference "CBD". The characteristic of the personal CBD tweets are those tweets that appear to be authored by an individual that is making a reference to their use of CBD. Whereas the characteristic of the commercial CBD tweets are those that indicate sales and/or promotion of CBD and may be authored by individual or organizations.

**Table 1.** Example personal and commercial CBD-related tweets.

| Personal CBD | Commercial CBD | Non-personal/non-commercial CBD |
|---|---|---|
| CBD products are soooo good for anxiety and they don't make you high | Pain! Pain! Go Away!!!! We have a variety of CBD products for your needs. Make sure to ask about our selection your next visit. URL | If you live where medical marijuana is legal, you could get paid $3k a month to critique weed, CBD, edibles and more URL |
| I've used CBD for anxiety. It is WAY healthier than taking benzodiazepines all the time, which is the only other thing that helps with it. I also use CBD for pain. You know what else is bad for your liver? Tylenol and Ibprofen | Over time, poor sleep can leave you feeling wrecked. Could CBD help? URL #cbd #cbdoil #hemp #cannabis #sleep #insomnia | The FDA is worried about CBD. Should you be concerned? URL |
| Take some painkillers with sleeping aid like Tylenol or Advil PM or something. If you got any CBD or weed maybe try that too | Chronic Fatigue and Cannabis CBD THC oil - URL | This room is half the size of my cbd apartment I'm paying 2k a month for |
| CBD gummies will not give you the high, but for me personally CBD oil edibles helped with anxiety and menstrual cramps. It does affect people differently but its worth a shot to try | Our CBD cream combines the relief potential of arnica and natural menthol oil with cocoa butter and the scents of eucalyptus & lavender. This is relief that is crafted for all of your senses. URL | Flinders Street in Melbourne's CBD has been re-opened following an earlier protest. Traffic delays are easing rapidly. Tram services have also resumed through the area. Thanks for your patience during this disruption. #victraffic |

Before training the binary classifiers to sort the full dataset into personal and commercial CBD-related tweets, we preprocessed the text of the tweets by normalizing all URLs to one consistent string, removing special characters and English stop words, lemmatizing the tweet text, and converting all of the text to lowercase. The binary classifiers were trained on 80% of this annotated sample and tested on the remaining 20% of the annotated sample. We created a matrix of the term frequency–inverse document frequency (TF-IDF) features based on the words within tweets using a range of n-grams from one to three, as well as a matrix of the TF-IDF features based on the characters within tweets using a range of n-grams from three to six. The resulting matrices were stacked horizontally, which served as the input to our model for training the classifiers.

**Table 2.** Training set personal CBD class counts

|                  | Pre-downsampling | Post-downsampling |
|------------------|------------------|-------------------|
| Personal CBD     | 631              | 631               |
| Non-personal CBD | 4,865            | 631               |
| Total            | 5,496            | 1,262             |

**Table 3.** Training set commercial CBD class counts

|                    | Pre-downsampling | Post-downsampling |
|--------------------|------------------|-------------------|
| Commercial CBD     | 489              | 489               |
| Non-commercial CBD | 5,007            | 489               |
| Total              | 5,496            | 978               |

An analysis of the manually annotated tweets indicated that the classes of personal and commercial CBD-related datasets were imbalanced; the non-personal CBD-related tweets occurred 7.7 times more than the personal CBD related tweets, while the non-commercial CBD-related tweets occurred 10.2 times more than the commercial CBD-related to tweets. In order to achieve a balance of the classes in the training sets, we down-sampled both of the negative classes (non-personal/non-commercial CBD related) in the training set by taking a random set equivalent in size to the positive class (personal/commercial CBD related). Tables 2 and 3 show the class frequencies for both the personal and commercial CBD-related tweet classes prior to and post downsampling.

To train the two binary classifier we performed a grid search using a linear support vector classifier, logistic regression, Gaussian Naive Bayes classifier, and a random forest classifier to find the optimal classification algorithm and combination of parameters on our training sample. The range of parameters are found in the Table 4. After training the binary classifiers we applied each model to the larger CBD corpora of tweets.

We applied each of the classification algorithms on the corpora of tweets. Table 5 shows the amount of tweets that each classifier predicted. The Gaussian Naive Bayes predicted the fewest personal CBD tweets (169,876), whereas the random forest predicted the most personal CBD tweets (166,657), a difference of 10,577 tweets. The Gaussian Naive Bayes predicted the most commercial CBD tweets (200,473), whereas the random forest predicted the fewest commercial CBD tweets (92,524), a difference of 107,949 tweets.

**Table 4.** Algorithm & parameters combinations used in text classification tuning

| Algorithm | Parameter | Range |
|---|---|---|
| Linear support vector | Penalty | {l1,l2} |
| | Loss | {hinge , squared hinge} |
| | Regularization parameter | $X_k = 10^{a+(b-a)(k-1)/(n-1)}$, k=1,..,n; a=0; b=5; n=20 |
| LogisticRegression | Penalty | {l1,l2} |
| | Regularization parameter | $X_k = 10^{a+(b-a)(k-1)/(n-1)}$, k=1,..,n; a=0; b=5; n=20 |
| | Solver | {newton-cg,lbfgs,liblinear,ag,saga} |
| Gaussian Naive Bayes | Smoothing | {.99,.75,.5,.25,.1,.01,.001} |
| Random forest | Estimators | {2,5,10, 100, 1000} |
| | Maximum features | {1,2,4,8,12,20,30,40} |

**Table 5.** Counts predicted CBD tweets per binary classification algorithm

| | Gaussian NB | Logistic regression | Linear support vector | Random forest |
|---|---|---|---|---|
| Personal | 144,537 | 169,876 | 160,513 | 165,880 |
| Commercial | 195,483 | 148,866 | 150,431 | 92,168 |

## 4    Results

We trained each binary classification algorithm, for both the personal and the commercial CBD tweet binary classifiers independently. Table 6 shows the performance of each of the personal CBD binary classifiers and Fig. 2 shows the ROC curve and AUC scores. The Logistic Regression (penalty = l1, regularization = 69.52, solver = saga) binary classifier model provided the best results of the four binary personal CBD classifiers with an average F1-score and AUC score of 0.85. When this binary classifier was applied to the collection of tweets, it classified 169,876 tweets as personal CBD-related. Table 6 also shows the performance of each of the commercial CBD binary classifiers and Fig. 3 shows the ROC curve and AUC scores. The Logistic Regression (penalty = l1, regularization = 428.13, solver = saga) binary classifier model provided the best results of the four binary commercial CBD classifiers with an average F1-score and AUC score of 0.88. When this binary classifier was applied to the collection of tweets, it classified 148,866 tweets as commercial CBD-related.

**Table 6.** Personal/commercial CBD binary classifier performance metrics (20% test set results)

| | GaussianNB | | | | Logistic regression | | | | Linear support vector | | | | Random forest | | | |
|---|---|---|---|---|---|---|---|---|---|---|---|---|---|---|---|---|
| | Prec | Rec | F1 | Sup | Prec | Rec | F1 | Sup | Prec | Rec | F1 | Sup | Prec | Rec | F1 | Sup |
| Non-Pers CBD | 0.90 | 0.80 | 0.85 | 138 | 0.93 | 0.78 | 0.85 | 138 | 0.92 | 0.78 | 0.84 | 138 | 0.83 | 0.74 | 0.78 | 138 |
| Pers CBD | 0.79 | 0.90 | 0.84 | 115 | 0.78 | 0.93 | 0.85 | 115 | 0.78 | 0.92 | 0.84 | 115 | 0.72 | 0.82 | 0.77 | 115 |
| Average | 0.85 | 0.85 | 0.85 | | 0.86 | 0.86 | 0.85 | | 0.85 | 0.85 | 0.84 | | 0.78 | 0.78 | 0.78 | |
| | GaussianNB | | | | Logistic regression | | | | Linear support vector | | | | Random forest | | | |
| | Prec | Rec | F1 | Sup | Prec | Rec | F1 | Sup | Prec | Rec | F1 | Sup | Prec | Rec | F1 | Sup |
| Non-Com CBD | 0.92 | 0.77 | 0.84 | 95 | 0.91 | 0.83 | 0.87 | 95 | 0.92 | 0.82 | 0.87 | 95 | 0.78 | 0.92 | 0.84 | 95 |
| Com CBD | 0.81 | 0.94 | 0.87 | 101 | 0.85 | 0.92 | 0.89 | 101 | 0.85 | 0.93 | 0.89 | 101 | 0.90 | 0.75 | 0.82 | 101 |
| Average | 0.87 | 0.86 | 0.86 | | 0.88 | 0.88 | 0.88 | | 0.89 | 0.88 | 0.88 | | 0.84 | 0.84 | 0.83 | |

Since our test set in Table 6 was so small, we conducted a post-classification test in order to provide assurance that our binary classifiers were performing consistently. We accomplished this by taking a random set of 500 unique tweets from our collection, manually labeling them as we did for our initial training and test, and ran each tweet through the each of the four binary classifiers. Table 7 shows the result of this post-classification test. We observed that the logistic regression models for both the commercial and personal tweets provided the best results, when compared to the other three algorithms.

**Table 7.** Personal/commercial CBD binary classifier performance metrics (post-classification random sample, n = 500)

| | GaussianNB | | | | Logistic regression | | | | Linear support vector | | | | Random forest | | | |
|---|---|---|---|---|---|---|---|---|---|---|---|---|---|---|---|---|
| | Prec | Rec | F1 | Sup | Prec | Rec | F1 | Sup | Prec | Rec | F1 | Sup | Prec | Rec | F1 | Sup |
| Non-Pers CBD | 0.92 | 0.94 | 0.93 | 367 | 0.94 | 0.91 | 0.93 | 367 | 0.93 | 0.92 | 0.92 | 367 | 0.92 | 0.90 | 0.91 | 367 |
| Pers CBD | 0.82 | 0.77 | 0.80 | 133 | 0.78 | 0.85 | 0.81 | 133 | 0.79 | 0.80 | 0.79 | 133 | 0.75 | 0.79 | 0.77 | 133 |
| Average | 0.87 | 0.86 | 0.87 | | 0.86 | 0.88 | 0.87 | | 0.86 | 0.86 | 0.86 | | 0.84 | 0.85 | 0.84 | |
| | GaussianNB | | | | Logistic regression | | | | Linear support vector | | | | Random forest | | | |
| | Prec | Rec | F1 | Sup | Prec | Rec | F1 | Sup | Prec | Rec | F1 | Sup | Prec | Rec | F1 | Sup |
| Non-Com CBD | 0.92 | 0.83 | 0.87 | 367 | 0.91 | 0.93 | 0.92 | 367 | 0.91 | 0.94 | 0.92 | 367 | 0.86 | 0.98 | 0.91 | 367 |
| Com CBD | 0.62 | 0.79 | 0.70 | 133 | 0.79 | 0.74 | 0.77 | 133 | 0.81 | 0.74 | 0.77 | 133 | 0.89 | 0.55 | 0.68 | 133 |
| Average | 0.77 | 0.81 | 0.79 | | 0.85 | 0.84 | 0.85 | | 0.86 | 0.84 | 0.85 | | 0.88 | 0.77 | 0.80 | |

Having described the application of the classifiers, we can turn to term analysis. We generated unigram frequencies for both the personal and commercial corpora of tweets. We looked at the top 1,000 occurring terms (minus common English stopwords) and manually checked if the term was relevant to health, wellness, diseases, side effects, conditions, body parts, and/or references to other substances against standard English, medical, and slang dictionaries.

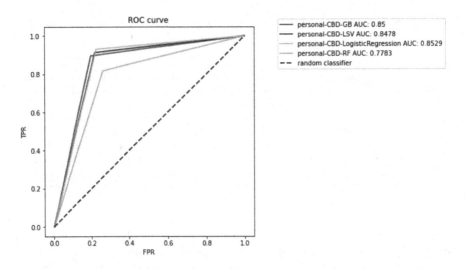

**Fig. 2.** Personal CBD binary classifier ROC/AUC

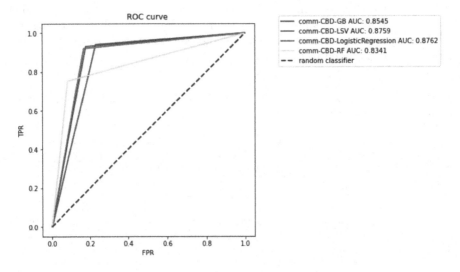

**Fig. 3.** Commercial CBD binary classifier ROC/AUC

We categorized the terms into 3 groups: health/medical, cannabis-related terms, and other substances. Within the health/medical terms, we included terms related to diseases, aliments, symptoms, body parts. We grouped cannabis-related terms together and separated them from the other substances groups since there seemed to be an overlap of CBD and THC-related tweets, both of which fall under the broader cannabis plant. The other substances groups include terms that refer to any other drug or medication. There were a few instances of words being included in multiple groups. For instance, high is a side effect and a term commonly used within cannabis. We used the Scattertext Python package to generate Figs. 4, 5 and 6 which provides a graphical representation of the frequencies within the personal and commercial CBD classes for each of the three term groups[4].

Within the other tweets making cannabis references, it seems that THC-related terms are mentioned within both personal and commercial corpora of tweets, with more hashtags containing these references more frequently in the commercial CBD tweets. The terms drink, melatonin, and pills were mentioned in the other substance set of tweets in both the personal and commercial CBD tweets. Kratom and MCT were mentioned more frequently within the commercial CBD tweets and less frequently in the personal CBD tweets. References to alcohol appear to occur slightly more than average within the personal CBD tweets and below average in the commercial CBD tweets. Opioids were mentioned but infrequently in both personal and commercial CBD tweet classes. We can see that, in terms of the health and wellness related terms, pain, sleep, and anxiety occurred frequently in both the personal and commercial CBD classes. We can see terms related to fitness and nutrition are found more frequently within the commercial CBD tweets. Tweets referencing PTSD appear to occur average within both classes. CBD tweets referencing autism appear to occur more than average within personal tweets, but infrequently within commercial tweets.

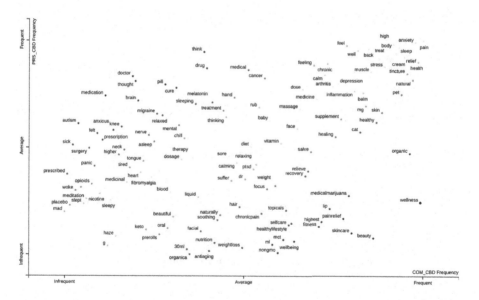

**Fig. 4.** Medical/health/wellness-related term frequency per class

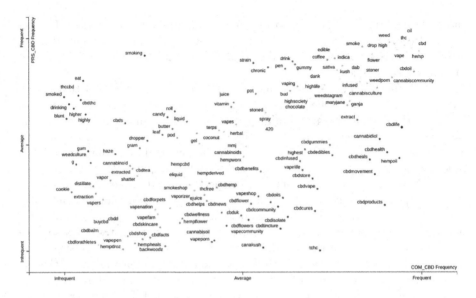

**Fig. 5.** Cannabis-related term frequency per class

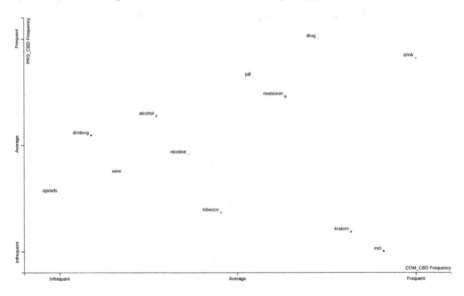

**Fig. 6.** Other substances term frequency per class

## 5 Conclusion

Text classification of tweets provides a mean to segment tweets into defined groups at a large scale. We have demonstrated that we can do this with tweets related to CBD by using text classification to identify tweets that reflect personal usage of CBD and tweets that reflect sales and/or commercialization of CBD. This classification work is an important example of data science in application, as CBD has not been subjected to the same tests and clinical trials as modern medications, yet is currently being used to treat a variety of conditions without proof of safety or efficacy. Our analysis provides a methodology to identify the medical terms of interest that are frequently being referenced within the commercial and personal corpora of CBD tweets, as well as a comparison of these term frequencies in relation to the document class (commercial or personal). This provided us the ability to see medical conditions that are commonly referenced in both document classes at high frequencies, as well as terms that are occurring more frequently in one document class over the other. When we applied these two classifiers to the collection of tweets, we identified multiple medical conditions, body parts, symptoms, other substances, and cannabis references that were mentioned at high frequencies in both personal and commercial CBD corpora as well as conditions that were mentioned disproportionately in one corpora over the other. Our results agree with previous research in that we also observed that CBD is associated with anxiety and pain; however, it also provided the ability to identify the terms associated with the personal use of CBD versus the terms associated with online marketing, in addition to the general medical terminology associated with CBD. In the future we will apply sentiment analysis to both the personal and commercial CBD tweets. This will allow us to gauge the satisfaction of CBD by those using it to treat specific conditions, along with providing the ability to compare the sentiment conveyed in the commercial CBD tweets versus the sentiment measured in the personal CBD tweets.

# References

1. Boyaji, S., Merkow, J., Elman, R.N.M., Kaye, A.D., Yong, R.J., Urman, R.D.: The role of cannabidiol (CBD) in chronic pain management: an assessment of current evidence. Curr. Pain Headache Rep. **24**(2), 1–6 (2020). https://doi.org/10.1007/s11916-020-0835-4

2. Campos, A.C., Moreira, F.A., Gomes, F.V., Bel, E.A.D., Guimarães, F.S.: Multiple mechanisms involved in the large-spectrum therapeutic potential of cannabidiol in psychiatric disorders. Philos. Trans. R. Soc. B Biol. Sci. **367**(1607), 3364–3378 (2012). https://doi.org/10.1098/rstb.2011.0389

3. Johnson, J.R., Burnell-Nugent, M., Lossignol, D., Ganae-Motan, E.D., Potts, R., Fallon, M.T.: Multicenter, double-blind, randomized, placebo-controlled, parallel-group study of the efficacy, safety, and tolerability of THC: CBD extract and THC extract in patients with intractable cancer-related pain. J. Pain Symptom Manag. **39**(2), 167–179 (2010). https://doi.org/10.1016/j.jpainsymman.2009.06.008

4. Kessler, J.: Scattertext: a browser-based tool for visualizing how corpora differ. In: Proceedings of ACL 2017, System Demonstrations. Association for Computational Linguistics (2017). https://doi.org/10.18653/v1/p17-4015

5. Kosgodage, U.S., et al.: Cannabidiol (CBD) is a novel inhibitor for exosome and microvesicle (EMV) release in cancer. Front. Pharmacol **9** (2018). https://doi.org/10.3389/fphar.2018.00889

6. Linares, I.M.P., et al.: No acute effects of cannabidiol on the sleep-wake cycle of healthy subjects: a randomized, double-blind, placebo-controlled, crossover study. Front. Pharmacol. **9** (2018). https://doi.org/10.3389/fphar.2018.00315
7. Masataka, N.: Anxiolytic effects of repeated cannabidiol treatment in teenagers with social anxiety disorders. Front. Psychol. **10** (2019). https://doi.org/10.3389/fpsyg.2019.02466
8. Narayanan, S., et al.: Cannabidiol (CBD) oil, cancer, and symptom management: a google trends analysis of public interest. J. Altern. Complement. Med. **26**(4), 346–348 (2020). https://doi.org/10.1089/acm.2019.0428
9. Palmieri, B., Laurino, B., Vadalà, M.: A therapeutic effect of CBD-enriched ointment in inflammatory skin diseases and cutaneous scars. LA CLINICA TERAPEUTICA **170**(2), 93–99 (2019). https://doi.org/10.7417/CT.2019.2116
10. Shannon, S.: Cannabidiol in anxiety and sleep: a large case series. Permanente J. (2019). https://doi.org/10.7812/tpp/18-041
11. Skelley, J.W., Deas, C.M., Curren, Z., Ennis, J.: Use of cannabidiol in anxiety and anxiety-related disorders. J. Am. Pharm. Assoc. **60**(1), 253–261 (2020). https://doi.org/10.1016/j.japh.2019.11.008
12. Aggarwal, S.K., et al.: Medicinal use of cannabis in the united states: historical perspectives, current trends, and future directions. J. Opioid Manag. **5**(3), 153 (2018). https://doi.org/10.5055/jom.2009.0016
13. Tran, T., Kavuluru, R.: Social media surveillance for perceived therapeutic effects of cannabidiol (CBD) products. Int. J. Drug Policy **77** (2020). https://doi.org/10.1016/j.drugpo.2020.102688
14. Überall, M.A.: A review of scientific evidence for THC:CBD oromucosal spray (nabiximols) in the management of chronic pain. J. Pain Res. **13**, 399–410 (2020). https://doi.org/10.2147/jpr.s240011
15. U.S. Food and Drug Administration: FDA and Cannabis: Research and Drug Approval Process. https://www.fda.gov/news-events/public-health-focus/fda-and-cannabis-research-and-drug-approval-process

# Finding Records in Social Media: A Natural Language Processing Fundamentals Exploration

Babatunde Kazeem Oladejo[1]([✉]) [iD], Sunčica Hadžidedić[2] [iD], and Emir Ganić[1]

[1] Department of Computer Science and Information Systems,
University Sarajevo School of Science and Technology, 71210 Ilidza, Bosnia and Herzegovina
babatunde.oladejo@stu.ssst.edu.ba, emir.ganic@ssst.edu.ba
[2] Department of Computer Science, Durham University, South Road, Durham D1 3LE, UK
suncica.hadzidedic@durham.ac.uk

**Abstract.** Social media postings are now routinely used as proof of activities, events, or transactions in news media, academic institutions, governments, judicial courts, commerce, and various other organizations. The need to preserve social media content as records has drawn the interest of academic researchers, industry professionals, and policy makers. Despite the importance of this research area, selection of records from a pool of social media content remains an area of low research activity. This paper explores the use of Natural Language Processing methods to classify and select records from a pool of tweets (twitter social media content). We experiment with various characteristics of the data and NLP parameters with the goal of determining optimal parameters for training a supervised machine learning classifier. This paper can serve as an aid for understanding the fundamental elements of automating the selection of social media records.

**Keywords:** Social media record · Record selection · Record classification · Machine learning · NLP

## 1 Introduction

When Darnella Frazier, a teenager from Minneapolis, MN, USA decided to post a video of George Floyd's police brutality on her Facebook social media page, she did not realize how big the impact would be. Her attorney, Seth Cobin said in the Star Tribune [1] "She had no idea she would witness and document one of the most important and high-profile police murders in American history". The social media post went viral, bypassing mass media, crime reporting agencies and all traditional record capture to become the central evidence against the police officers charged in the court case.

Beyond the law and order example above, the use of records from social media has become prevalent in education [2], mass media [3], medical practice [4], government administration [5], corporate business [6], and other human-cultural organizations. This engaged utilization of records from social media prompts the questions: what is a social media record? What are the criteria for the selection of social media records from a pool of social media content?

© Springer Nature Switzerland AG 2021
J. Hasic Telalovic and M. Kantardzic (Eds.): MeFDATA 2020, CCIS 1343, pp. 151–164, 2021.
https://doi.org/10.1007/978-3-030-72805-2_11

**What is a Social Media Record?**

The US National Archives and Records Administration (NARA) in its Bulletin 2014-02 [7] defined social media records as social media content that fulfils the record criteria of material that is recorded, made or received in the course of official business, regardless of its form or characteristics, and is worthy of preservation. A social media record must have content, context, and structure along with associated metadata (e.g., author, date of creation), and be properly maintained to ensure reliability and authenticity. To assist agencies in determining the record status of social media content, NARA further specified a (non-exhaustive) list of qualification questions:

- Does it contain evidence of an agency's policies, business, or mission?
- Is the information only available on the social media site?
- Does the agency use the tool to convey official agency information?
- Is there a business need for the information?

A 'yes' answer to any of the questions would make the social media content selectable as a record.

Although developed for US federal agencies, the NARA social media record policy addresses several of the issues and challenges facing other organizations [8]. An establishment's record policy usually dictates its record selection and records classification agenda. Records selection and records classification are fundamental processes of Records Management [9] and are essential to any study in a new and developing subsection, such as social media. This paper will focus on these two interwoven elements in our exploration of social media records management.

Finding social media records in a pool of social media content is a challenging problem to solve because, practically, any content can be a record [10]. For example, a casual "hello" social contact between two people in a criminal investigation might become a record if both parties claim not to know each other. However, such a casual "hello" social contact between ordinary people would not be considered records.

For this research, instead of attempting to directly decipher social media records from a pool of social media content, we took the approach of using an already curated record source and subsequently examine the characteristics of the records. A successful exploration of the data would enable us to use Natural Language Processing (NLP) methods to elicit features from the textual content of the records. These features could then be used to train a machine learning classifier to identify new records in a new pool of social media content. Our chosen dataset is twitter social media postings in the news publications of the British Broadcasting Corporation (BBC) and the New York Times (NYT). Articles published by news publishers are often referred to as credible sources of information curated for the public under due editorial processes, and therefore acceptable as records [11].

With the foregoing approach to the identification of the social media record and its management, we articulate our research goals as:

1. Explore the selected social media record dataset for its properties, which can assist in the implementation of an automated record selection initiative.

2. Explore the use of fundamental NLP techniques to classify records in the selected social media dataset and determine the best set of parameters and algorithms for the training of a machine learning classifier.

The rest of this paper is organized as follows. Section 2 discusses the literature related to the research objectives. Section 3 covers data collection and characteristics of the dataset, while Sect. 4 details the pre-processing work done on the dataset. In Sect. 5, we annotate the experiments performed and discuss the results. Section 6 summarizes the research and discuss potential areas for future improvement and limitations.

## 2 Related Works

Our search of academic databases namely, Google Scholar, DeepDyve and Springer Link for articles related to Social Media Records Management returned only about 200 articles. Several of the articles belong to the archivist, records management and law domains and only describe the nature of the social media record from theoretical and application practice viewpoints [12–14]. We found fewer articles that describe the computing nature of social media records, which is the primary focus of our research. Van Wyk and Starbird [15] used the term social media record to describe a collection of messages in a time-framed event such as an earthquake. The idea of the social media record being a collection of related messages was also used by Liu et al. [16], but with a patient's social media message collection as the record. The researchers created a custom algorithm called SocInf to evaluate their hypothesis of Membership Inference attacks by potential hackers. SocInf's performance was compared against three other machine learning models trained on logistics regression, Xgboost and BigML, a cloud-based platform. SocInf was found to be the best performer in their experiments.

Given the lack of computing research materials on social media records management, we considered studies on social media topic classification that use NLP methods and approaches in line with our research goals. Topic classification deals with the grouping of social media content [17] and is well aligned with record classification, where record is a subset of content [7]. We narrowed down the vast number of search results to articles that address news related topical classifications of social media content in line with our research dataset and experiment objectives and report their findings in the following text.

### Dataset Features and Pre-processing
Iman, Zahra, et al. [17] conducted a large longitudinal study of twitter topic classification of using over 800 million English language tweets between 2013 and 2014. The study found that Naïve Bayes is an effective topical learner which could generalize and generate new previously unseen news-worthy topics after a year-long training. The authors compared the effectiveness of various features and found hashtags, mentions, locations to be amongst the best features for training their classifier.

Perreault and Ruths [18] found that manual labelling for supervised topic classification result in higher accuracy and precision than unsupervised methods, but labelling can be labour intensive, time-consuming, and expensive. Semi-automated labelling can also achieve good results given adequate use of training instances [19]. To improve the quality of social media content for classification, pre-processing is essential. Pre-processing, textual clean-up or text massaging tasks often include removal of stop-words, part of speech (POS) tagging and replacement of non-standard words (such as "lmao", "cuz", "lol", etc. with the standard equivalents) [20].

**NLP Algorithms and Performance Measurements**

Working with a team of journalists to identify newsworthy events that were likely to become rumours, Zubiaga et al. [21] used a linear-chain Conditional Random Fields (CRF) algorithm to learn the dynamics of information during breaking news and classified the information as rumour or non-rumour. A performance comparison between CRF and other classifiers (SVM, Random Forest, Naïve Bayes, and Maximum Entropy) was conducted. While SVM best exploited the social (twitter) features, CRF was better with source (news publisher) features. The overall result showed that CRF performed best in terms of precision but lacked behind in recall. The study concluded that when all constraints were considered, CRF outperformed the other classifiers in the detection of rumours.

A similar comparative evaluation of ML algorithms for the detection of credible news was conducted by Hassan et. al. [22]. The researchers compared 5 algorithms, namely, Linear Support Vector Machines (LSVM), Logistic Regression (LR), Random Forests (RF), Naïve Bayes (NB) and K-Nearest Neighbours (KNN). They found that the best performance was achieved with LSVM using a combination of unigrams and bigrams as features prioritized by TF-IDF (Term Frequency – Inverse Document Frequency).

The challenge of securing adequate data for supervised machine learning of fake news was addressed by Helmstetter and Paulheim [23] by using a technique called weakly supervised learning. A dataset of tweets was automatically labelled by the source reliability, i.e. trustworthy or untrustworthy source, and a classifier trained on the dataset. The classifier was then repurposed for a different classification target, i.e., the classification of fake and non-fake tweets. Interestingly, the labels were not always accurate according to the new classification target (i.e. not all tweets by an untrustworthy source turned out to be fake news, and vice versa), the research show that despite the inaccuracy of the original dataset, fake news could be detected with an F1 score of up to 0.9 using the XGBoost classifier.

Salminen, Joni, et al. [24] created a detailed taxonomy of online hate types and people targeted as part of an effort to automate the detection of online hate expressions. The researcher created ML models that classifies the hateful comments, experimenting with Logistic Regression, Decision Tree, Random Forest, Adaboost, and SVM. The study found that SVM performed the best for the dataset, with an average F1 score of 0.79.

Overall, this review of literature related to our research goals highlights the importance of clarity of the problem to be solved and collection of relevant data including proper labels for supervised machine learning. Additionally, adopting good preprocessing and feature selection strategies are central to all the implementations. Lastly, we found that the most compared algorithms for news topic classification are XGBoost, SVM, Naïve Bayes and Random Forest, and we intend to benchmark these in our experiments.

# 3   Dataset

Over the period of four months (October 2019 and January 2020), we scraped the websites of the British Broadcasting Corporation (BBC) and New York Times (NYT) for articles that include twitter URLs (Unique Resource Locators), irrespective of topic area. Our Python BeautifulSoup API based custom scrapper retrieved a total of 5,305 Record Tweets (RecTweets) from the news websites. This dataset is called the News-cited dataset. Using the TWARC utilities from the DocumentingNow project [25], the tweet IDs from the News-cited dataset were hydrated to JSONL format. Another python program was used in conjunction with the TWARC-Replies utility to retrieve over 2 million replies or Supporting Tweets (SupTweets) from the Twitter Search API stream.

Initial exploratory data analysis was performed on the data with the following outcome:

- Total 5,305 RecTweets extracted from BBC and NYT

  - 5,041 successfully hydrated from Twitter (264 could not be found - might have been deleted by the owners).
  - 4,708 were English language tweets.
  - 1,583 of the English language RecTweets were properly pre-classified into content categories by the news publishers.

- Total 2,429,549 SupTweets retrieved from Twitter Standard Search API[1] / TWARC-Replies

  - 1,980,781 were English language tweets
  - 4,451 of the 4,708 English language RecTweets had at least 1 SupTweet
  - Large variance observed in the number of SupTweets per RecTweets (from 0 to > 40,000). See Fig. 1.

- Total 14 topical categories were provided from the news publishers (see: Table 1). The categories will be used as the class labels for the supervised machine learning models in this research.

[1]This research used the free Standard Search API, which provides free access to public tweets posted within the past 7 days only [26]. This restriction contributes to the unavailability of some SupTweets replies in our dataset.

# 4   Pre-processing

The tweet text was cleaned up by removing all special characters, URLs, html codes, emojis, hashtags, and user mentions. The emojis, hashtags and user mentions were preserved in separate fields. Hashtags and mentions were de-duplicated to ensure uniqueness of the stored values for the cases where repeated in the RecTweets and SupTweets. All unicode characters were converted to ASCII using the unidecode Python module. For

**Table 1.** Topical categories

| ID | Class | All content | Publisher classified, english language |
|----|-------|-------------|----------------------------------------|
| 1 | Arts | 14 | 11 |
| 2 | Books | 23 | 20 |
| 3 | Business | 139 | 119 |
| 4 | Climate | 12 | 12 |
| 5 | Economy | 12 | 12 |
| 6 | Entertainment | 464 | 429 |
| 7 | Fashion | 3 | 3 |
| 8 | Food | 12 | 12 |
| 9 | Health | 14 | 14 |
| 10 | Politics | 715 | 685 |
| 11 | Sports | 172 | 152 |
| 12 | Technology | 120 | 112 |
| 13 | Travel | 2 | 2 |
| 14 | Unknown | 3,603 | – |
| | TOTAL | 5,305 | 1,583 |

example, a word like Tánaiste was converted to Tanaiste. The NLTK module was used to tokenize the tweet text, perform stemming (PorterStemmer) and lemmatization (Word-Net). We extended the standard NLTK stop-words list with an additional list of 290 + stop-words that are irrelevant to the classification task.

The pre-processing steps were done separately for RecTweets and SupTweets. All the replies (SupTweets) to a RecTweet were concatenated together to form a continues corpus of text. The RecTweet + SupTweets text is referred to as the Social Media Record (SMR) dataset in this paper and will be used for the NLP classification task.

**SupTweets Threshold**

To avoid overfitting and/or underfitting due to the large variance observed in the SupTweets, we limited the SMR dataset to SupTweets with minimum 10 replies and maximum 260 (corresponding to the median and upper whisker of the BoxPlot in Fig. 2, respectively). Additionally, to ensure that the most relevant tweets were included in the SMR dataset, given the cap of 260 replies, we applied the following sort order to the SupTweet stream for each RecTweet:

- SupTweet size >= 141, no URLs or Media, ordered by parsed_created_at
- SupTweet size <= 140, no URLs or Media, ordered by parsed_created_at
- SupTweet with URLs and Media (potential spams)

We consider a greater than 140 characters sized SupTweet a large sized tweet given that it was the old limit set by Twitter [27] before the recent revision to 280 characters. This scheme allowed us to capture the more relevant supporting tweets before reaching the maximum threshold of 260 SupTweets.

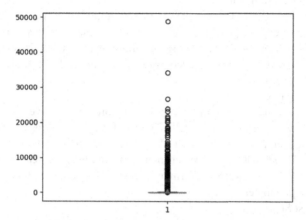

**Fig. 1.** Supporting Tweets (Replies) with outliers

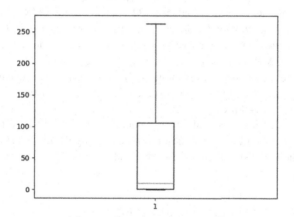

**Fig. 2.** Supporting Tweets (Replies) without outliers

## 5  Experiments

**Algorithms**
Based on our literature review of related works, we selected the most benchmarked machine learning classifiers for our experiments. These are: Linear Support Vector Machine (LSVM), Gaussian Naïve Bayes (GNB), Random Forest (RF) and XGBoost.

**Baseline Vectorizer**
The pre-processed SMR dataset was converted to a unigram Document Text Matrix

(DTM) using the Python sklearn CountVectorizer library as a baseline for the experiments. A preliminary review of the DTM revealed term sparsity issues. To reduce the sparsity, the following steps were taken:

1. Constant features removal
   Constant features are terms in the DTM that are found in only a document (SMR). These constant features provide little or no value to the classification task [28] and can be removed from the DTM. Setting the minimum Document Frequency (min_df) parameter of the vectorizer to 2, accomplishes the removal of the constant features while creating the DTM.
2. Maximum features
   It is good practise to set a maximum number of features for the vectorizer to use as a dimensionality reduction solution [29]. The number typical varies from 1,000 [30] to 20,000 [29] depending on the dataset. We set the baseline vectorizer's max_features parameter to 20,000. During the experiments, we vary the max_features parameter to 10,000 and 5,000 and observe performance metric changes to determine the optimum value for the parameter.
3. Training/Test Split
   For the baseline machine learning experiment, we divided the SMR dataset into training and test dataset using a 80% and 20% respective split ratio [31]. We encountered some challenges with class imbalance [34], where some SMR records with low counts in the dataset did not divide well with the stratify parameter of the Vectorizer. To resolve the problem, we removed records where there are less than 20 items in a class from the SMR dataset. This reduced the classes available in the SMR dataset to 5 (see Table 2). Despite this resolution, the observed class imbalance remained. Three of the classes had less than 100 records and the other two classes had over 250 records. A research decision to made to proceed with the class imbalance as further data reduction might affect the machine learning model efficacy. Finally, we performed a 5-fold cross validation on the training data. We chose 5-fold cross validation due to the small size of the data [32].

**Table 2.** Final SMR dataset classes

| ID | Class | SMR count |
|----|-------|-----------|
| 3 | Business | 67 |
| 6 | Entertainment | 264 |
| 10 | Politics | 389 |
| 11 | Sports | 97 |
| 12 | Technology | 57 |
|    | TOTAL | 874 |

**Experimental Results**

We ran a total of 23 experiments with various vectorizer parameters and feature sets. We executed 9 baseline experiments using the Baseline Vectorizer described in the subsection above and various elements of the SMR dataset. Experiment id "A8. Mentions + Nouns" produced the best baseline result with F1 scores ranging between 76% to 81% across all the 4 classifiers. The other baseline experiments produced metric scores ranging from 47% to 81%. We decided to carry forward the configuration for A8. as the baseline. For the simplicity of this report, we discuss the performance metrics in terms of F1 scores because it is considered the most comprehensive of the performance metrics [33] and derived from a combination of Precision and Recall, with the formula:

$$\text{F1 score} = 2 * ((\text{Precision} * \text{Recall})/(\text{Precision} + \text{Recall})) \tag{1}$$

In the next series of experiments, we varied n-grams on the baseline and found that both "B1. 1-g + 2-g" and "B2. 1-g + 2-g + 3-g" performed well with F1 score ranges of 76% to 81% across all the 4 classifiers. Without a clear winner in this test, we kept the A8. Configuration, but replaced the Count Vectorizer with a TF-IDF Vectorizer. The TF-IDF test produced F1 score ranges of 76% to 82%, which is a slight improvement, but not enough to justify replacing the baseline for the next set of experiments.

Using the baseline Count Vectorizer, we changed the max_features parameter to experiment with 5,000 and 10,000 Top-N features. The results remained flat with F1 score ranges of 76% to 81%.

For the final set of tests, we replaced the Count Vectorizer with a TF-IDF Vectorizer and varied Top-N and n-Gram features. This time, we got a decisive improvement with the Linear SVM algorithm having an F1 score of 84.97% in the "F2. TFIDF 10k Features 1ng_2ng" test. The other performance measurement: accuracy, precision and recall scores for the Linear SVM (Mentions + Nouns, TF-IDF, 10,000 max_features, ngram = 1, 2) test were also above the other algorithms in the test group and the baseline test (see Table 3).

**Table 3.** Baseline vs best result.

| Exp ID | Classifier | Exec time | Accuracy | F1 score | Precision | Recall |
|--------|-----------|-----------|----------|----------|-----------|--------|
| A8 | RandomForest | 10.96 | 80.92 | 76.04 | 76.81 | 80.92 |
| A8 | LinearSVM | 18.25 | 81.98 | 80.43 | 84.25 | 81.98 |
| A8 | GaussianNB | 3.769 | 80.46 | 80.60 | 82.64 | 80.46 |
| A8 | XGBoost | 219.2 | 83.97 | 81.69 | 85.61 | 83.97 |
| F2 | RandomForest | 11.97 | 82.12 | 77.17 | 78.78 | 82.12 |
| F2 | **LinearSVM** | **39.82** | **86.84** | **84.98** | **87.95** | **86.84** |
| F2 | GaussianNB | 4.353 | 79.40 | 79.16 | 81.30 | 79.40 |
| F2 | XGBoost | 252.2 | 83.83 | 81.55 | 85.23 | 83.83 |

**Confusion Matrix Analysis**

Given that our experimental results were based on weighted averages of the performance metrics of all the classes, it was challenging to assess the impact of the observed class imbalance as the performance of the individual classes could not be determined. To address this concern, we created confusion matrices of the exemplary experiments: "A8. Mentions + Nouns" and "F2. TFIDF 10k Features 1ng_2ng" using only the selected Linear SVM algorithm (see Tables 4 and 5). All the records in the SMR dataset were used for the confusion matrices on a 5-fold cross-validation.

**Table 4.** Confusion matrix for experiment A8

|  |  | Predicted | | | | | |
|---|---|---|---|---|---|---|---|
| Actual | Class ID | 3 | 6 | 10 | 11 | 12 | Sum |
|  | 3 | **43** | 16 | 8 | 0 | 0 | **67** |
|  | 6 | 13 | **231** | 20 | 0 | 0 | **264** |
|  | 10 | 0 | 0 | **384** | 1 | 4 | **389** |
|  | 11 | 0 | 0 | 40 | **57** | 0 | **97** |
|  | 12 | 0 | 0 | 35 | 2 | **20** | **57** |
| Sum |  | 56 | 247 | 487 | 60 | 24 | 874 |

**Table 5.** Confusion Matrix for Experiment F2

|  |  | Predicted | | | | | |
|---|---|---|---|---|---|---|---|
| Actual | Class ID | 3 | 6 | 10 | 11 | 12 | Sum |
|  | 3 | **33** | 27 | 7 | 0 | 0 | **67** |
|  | 6 | 1 | **261** | 2 | 0 | 0 | **264** |
|  | 10 | 0 | 0 | **383** | 2 | 4 | **389** |
|  | 11 | 0 | 0 | 39 | **58** | 0 | **97** |
|  | 12 | 0 | 0 | 38 | 1 | **18** | **57** |
| Sum |  | 34 | 288 | 469 | 61 | 22 | 874 |

Although it is possible to exhaustively analyse a Confusion Matrix, we chose to simplify the analysis in this report by comparing only the true-positive values of the matrices (highlighted diagonal values in Tables 4 and 5) of the two exemplar experiments against the actual class values (see Bar Plot in Fig. 3). Experiments A8 and F2 produced nearly identical performance on classes 10-Politics, 11-Sports and 12-Technology. Experiment F2 performed better than A8 on class 6-Entertainment by a 30-point margin, but lower than A8 on class 3-Business by a 10-point margin. Both experiments A8 and F2 generally performed poorly on the classes with low SMR counts (less than 100). Based on the comparable performance of Experiment F2 against Actual on the higher SMR count (greater than 250) classes, we accept the overall result of Experiment F2 being the better model, but caution that more data and further experiments are required to assert this conclusion.

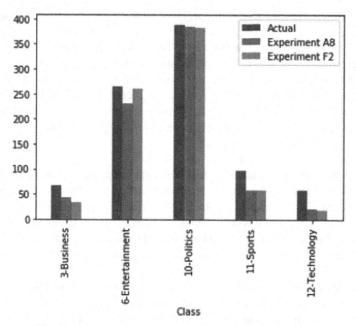

**Fig. 3.** Confusion matrix – actual vs. exemplar experiments

## 6  Discussions and Conclusion

We had two primary goals set for our research. The first was to explore the properties of a social media records dataset for characteristics that can assist in automatic record selection and secondly, to use NLP techniques to determine the best parameters and algorithms for the training of a machine learning classifier. To achieve these goals, we collected data from two reliable news sources: the BBC and NYT and experimented with 4 machine learning algorithms to automate the selection of records into record classifications.

For the **first goal**, we found that user mentions and nouns (names of people, places, and things), when combined, are the best natural language properties of the News-cited SMR dataset for record selection automation. We also found our scheme of assembling the social media record as a combination of the news-cited tweet and its replies, an effective lexical density strategy.

We fulfilled our **second goal** by performing 23 machine learning experiments using 4 different algorithms: Linear Support Vector Machine (LSVM), Gaussian Naïve Bayes (GNB), Random Forest (RF) and XGBoost. The results show that LSVM is the best performing algorithm with accuracy, F1, precision and recall scores of 86.84%, 84.98%, 87.95%, 86.84% respectively. While the SLVM program execution time of 39.87 s is not the fastest, it is much better than the slow running XGBoost time of 252.2 s.

**Future Work**

There are several areas we would like to consider for a future work on social media records classification. We would like to experiment with more data to see if higher number of classes and features will improve the performance of the machine learning model and correct the class imbalance observed in the current experimental dataset. Also, we would like to use social media data from sources beyond news articles to broaden the record selection capability of the model. Additionally, it would be interesting to experiment with other feature engineering techniques such as Mutual Information Gain and advanced machine learning algorithms, for example, Deep Learning.

**Limitations**

Our research has certain limitations. One significant issue is media posting often include images, videos, and links to external sources. These are excluded from the dataset used in this research, since we focused on text-based NLP techniques. It is also important to note that the result of this research is only a framework. While our research illustrates how a social media records classification can be developed and utilized in records selection, it is not comprehensive, and is limited to the few categories discovered in the news-cited dataset.

Overall, despite the limitations of our research work, we have laid a foundation for the automation of social media records selection and records classification, both of which are essential for a meaningful records management approach to social media. The big data nature of social media especially in terms of unstructured natural language variety makes the use of machine learning compelling, if not a must-have solution. We consider our recommendations of the social media record structure (as a main record plus supporting replies), feature selection (mentions and nouns) and algorithm (linear SVM) noteworthy milestones as we continue to work towards finding solutions that lead to a future, better managed social media ecosystem.

# References

1. Walsh, P.: Star Tribune, 11 June 2020. https://www.startribune.com/teen-who-shot-video-of-george-floyd-wasn-t-looking-to-be-a-hero-her-lawyer-says/571192352. Accessed 11 Sep 2020
2. Pondiwa, S., Phiri, M.: Challenges and opportunities of managing social media generated records in institutions of learning: a case of the Midlands State University, Zimbabwe. In: Tatnall, Arthur, Mavengere, Nicholas (eds.) SUZA 2019. IAICT, vol. 564, pp. 145–156. Springer, Cham (2019). https://doi.org/10.1007/978-3-030-28764-1_17
3. Zubiaga, A.: Mining social media for newsgathering: a review. Online Soc. Netw. Media **13**, 100049 (2019)
4. Eggleston, E.M., Weitzman, E.R.: Innovative uses of electronic health records and social media for public health surveillance. Curr. Diab. Rep. **14**(3), 1–9 (2014). https://doi.org/10.1007/s11892-013-0468-7

5. Bertot, J.: Social media, open platforms, and democracy: transparency enabler, slayer of democracy, both? In: Proceedings of the 52nd Hawaii International Conference on System Sciences (2019)
6. Franks, P., Doyle, A.: Retention and disposition in the cloud-do you really have control? In: Proceedings of the International Conference on Cloud Security Management ICCSM 2014 (2014)
7. NARA, National Archives and Records Administration: Guidance on managing social media records. https://www.archives.gov/records-mgmt/bulletins/2014/2014-02.html. Accessed 15 Sep 2020
8. Iron Mountain, Social Media Records call for fresh approach. https://www.ironmountain. com/resources/general-articles/s/social-media-records-call-for-fresh-approach. Accessed 17 Sep 2020
9. Cisco, S.L., Strong, K.V.: The value added information chain. Inf. Manag. **33**(1), 4 (1999)
10. Low, J.T.: A literature review: what exactly should we preserve? How scholars address this question and where is the gap. arXiv preprint arXiv:1112.1681 (2011)
11. Deacon, D.: Yesterday's papers and today's technology: digital newspaper archives and 'push button' content analysis. Eur. J. Commun. **22**(1), 7 (2007)
12. Caron, D., Brown, R.: Appraising content for value in the new world: establishing expedient documentary presence. Am. Arch. **76**(1), 135–173 (2013)
13. Streck, H., and Endowment Fund: Social networks and their impact on records and information management. ARMA International Educational Foundation, pp. 3–9 (2011)
14. Strutin, K.: Social media and the vanishing points of ethical and constitutional boundaries. Pace Law Rev. **31**(1), 227–290 (2011). Article no. 6
15. Van Wyk, H., Starbird, K.: Analyzing Social Media Data to Understand How Disaster-Affected Individuals Adapt to Disaster-Related Telecommunications Disruptions (2020)
16. Liu, G., et al.: SocInf: membership inference attacks on social media health data with machine learning. IEEE Trans. Comput. Soc. Syst. **6**(5), 907–921 (2019)
17. Iman, Z., et al.: A longitudinal study of topic classification on Twitter. In: Eleventh International AAAI Conference on Web and Social Media (2017)
18. Perreault, M., Ruths, D.: The effect of mobile platforms on Twitter content generation. In: Fifth International AAAI Conference on Weblogs and Social Media (2011)
19. Cohen, R., Ruths, D.: Classifying political orientation on Twitter: it's not easy! In: Seventh International AAAI Conference on Weblogs and Social Media (2013)
20. Shahzad, B., et al.: Discovery and classification of user interests on social media. Inf. Discov. Delivery **45**(3), 130–138 (2017)
21. Zubiaga, A., Liakata, M., Procter, R.: Learning reporting dynamics during breaking news for rumour detection in social media. arXiv preprint arXiv:1610.07363 (2016)
22. Hassan, N.Y., et al.: Credibility detection in Twitter using word N-gram analysis and supervised machine learning techniques. Int. J. Intell. Eng. Syst. **13**(1), 291–300 (2020)
23. Helmstetter, S., Paulheim, H.: Weakly supervised learning for fake news detection on Twitter. In: 2018 IEEE/ACM International Conference on Advances in Social Networks Analysis and Mining (ASONAM). IEEE (2018)
24. Salminen, J., et al.: Anatomy of online hate: developing a taxonomy and machine learning models for identifying and classifying hate in online news media. In: ICWSM (2020)
25. TWARC, Documenting The Now. https://github.com/DocNow/twarc. Accessed 26 Sep 2020
26. Twitter, Standard Search API. https://developer.twitter.com/en/docs/twitter-api/v1/tweets/search/overview/standard. Accessed 26 Sep 2020
27. Gligorić, K., Anderson, A., West, R.: How constraints affect content: the case of Twitter's switch from 140 to 280 characters. arXiv preprint arXiv:1804.02318. (2018)

28. Chouhan, A., Prabhune, A.: FIF: a NLP-based feature identification framework for data warehouses. In: 2019 IEEE/WIC/ACM International Conference on Web Intelligence (WI). IEEE (2019)

29. Gimpel, K., et al.: Part-of-speech tagging for Twitter: annotation, features, and experiments (2010)

30. Curiskis, S.A., et al.: An evaluation of document clustering and topic modelling in two online social networks: Twitter and Reddit. Inf. Process. Manag. **57**(2), 102034 (2020)

31. Abro, S., et al.: Aspect based sentimental analysis of hotel reviews: a comparative study. Sukkur IBA J. Comput. Math. Sci. **4**, 11–20 (2020)

32. Hollenstein, N., et al.: Advancing NLP with cognitive language processing signals. arXiv preprint arXiv:1904.02682 (2019)

33. Lee, L.-H., Yu, L.-C., Chang, L.-P.: Overview of the NLP-TEA 2015 shared task for Chinese grammatical error diagnosis. In: Proceedings of the 2nd Workshop on Natural Language Processing Techniques for Educational Applications (2015)

34. Luque, A., et al.: The impact of class imbalance in classification performance metrics based on the binary confusion matrix. Pattern Recogn. **91**, 216–231 (2019)

# Author Index

Printed in the United States
by Baker & Taylor Publisher Services